Jo Gillespie

Principles of Organometa

for

£2-75

Principles of Organometallic Chemistry

G. E. Coates
Professor of Chemistry in the University of Wyoming

M. L. H. Green
Fellow and Tutor in Inorganic Chemistry, Balliol College, Oxford

P. Powell
Lecturer in Chemistry, Royal Holloway College, University of London

K. Wade
Reader in Chemistry, University of Durham

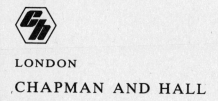

LONDON

CHAPMAN AND HALL

First published 1968
by Methuen & Co Ltd
Reprinted 1971
Reprint 1977 published by
Chapman and Hall Ltd
11 New Fetter Lane, London EC4P 4EE

© G. E. Coates, M. L. H. Green, P. Powell, K. Wade

Printed in Great Britain by
J. W. Arrowsmith Ltd., Bristol

ISBN 0 412 15350 5

Distributed in the U.S.A. by Halsted Press,
a Division of John Wiley & Sons, Inc., New York

Contents

4. ORGANOMETALLIC COMPOUNDS OF ELEMENTS OF MAIN GROUPS IV AND V

5. ORGANOMETALLIC COMPOUNDS OF THE *d*-BLOCK TRANSITION ELEMENTS: CLASSIFICATION OF LIGANDS AND THEORIES OF BONDING

6. PREPARATION OF ORGANO-TRANSITION METAL COMPOUNDS

7. REACTIONS AND STRUCTURES OF ORGANOMETALLIC COMPOUNDS OF THE TRANSITION ELEMENTS

8. THE ORGANIC CHEMISTRY OF FERROCENE AND RELATED COMPOUNDS

9. ORGANOMETALLIC COMPLEXES FORMED FROM ACETYLENES

10. THE ROLE OF ORGANOTRANSITION METAL COMPLEXES IN SOME CATALYTIC REACTIONS

Preface

The second edition of Organometallic Compounds (1960) was used not only by specialists but also as an undergraduate textbook. The third edition, recently published in two volumes, is about three times the length of the second and contains considerably more factual material than is appropriate for a student textbook. Therefore we believe that a shorter treatment would be welcome.

In planning this book the authors have emphasized matters more of principle than of detail, and have included in the first two chapters some general discussion of the properties and syntheses of organometallic compounds that is not to be found in the larger work. Some aspects of the organic chemistry of arsenic, and of silicon with particular reference to silicone polymers, are also included.

Most university teachers of chemistry are becoming seriously concerned about the relentless increase in the amount and complexity of the material that is squeezed into undergraduate chemistry courses. With this in mind the authors have tried to cut detail to a minimum, but readers will find that the relative amount presented varies considerably between the various topics discussed. In general the treatment is more extensive than usual only if either or both of these conditions are met: (1), the subject has significant bearing on other major branches of chemistry including important industrial processes; (2), the topic is commonly misunderstood or found to be confusing.

As most chemistry students know rather little about industrial processes, the relevance of organometallic chemistry to industry has been indicated and the scale of manufacture of some of the major industrial organometallics (lead antiknocks, silicones, and aluminium alkyls) has been emphasized. Thanks are due to the Continental Oil Company, Ethyl Corporation, and Imperial Chemical Industries for providing information in this connection.

Some of the most significant of all current chemical developments are in the area of catalysis, particularly by organometallic and transition metal compounds, and the book concludes with a short chapter on this subject dealing with the isomerization, oligomerization and oxidation of olefins, homogeneous hydrogenation, hydroformylation, and nitrogen fixation.

February 1968

G.E.C.
M.L.H.G.
P.P.
K.W.

NOTE ON THE 1971 REPRINT

We have taken the opportunity at this reprint of introducing into the text of 'Principles' new material dealing with some significant developments in organometallic chemistry that have occurred during the last two-and-a-half years. For instance, in chapter three, the section on carboranes has been extended and new synthetic applications of the hydroboration reaction have been noted.

Various errors have been eliminated and additional references have been incorporated in the lists at the end of each chapter.

G.E.C.
M.L.H.G.

October 1970

P.P
K.W.

CHAPTER ONE

General survey

Introduction

In this book we shall be concerned with the properties of compounds containing metal-carbon bonds.* Metal cyanides, carbonyls and carbides, although they contain metal-carbon bonds, and hence are organometallic compounds in terms of this definition, will not be discussed in detail here, since most of their chemistry is more usefully considered in conjunction with that of inorganic rather than organometallic compounds.

Historical background

Whereas some organometallic compounds have been known for a century or more, (e.g. the alkyls of zinc, mercury and arsenic), the development of the subject, especially of the organometallic chemistry of the transition elements, is much more recent. The study of organometallic compounds has often contributed significantly both to chemical theory and practice. Thus the preparation and investigation of the properties of ethylzinc iodide and of diethylzinc led Frankland (1853) to make the first clear statement of a theory of valency, in which he suggested that each element has a definite limiting combining capacity. More recently the chance synthesis of ferrocene (π-C$_5$H$_5$)$_2$Fe (1951) and the determination of its structure (see p 199) in the following year, opened up a field of research of hitherto unforeseen variety, which has contributed greatly to our understanding of chemical bonding. From the more practical standpoint, the discovery of the organomagnesium halides (Grignard reagents) in 1900 provided readily handled and versatile intermediates for a variety of organic and organometallic syntheses. An industrially applicable method for the preparation of organosilicon halides, which are intermediates in silicone polymer

---

*In the past many substances, such as metal alkoxides, which contain both metal and carbon atoms, but which lack a direct linkage between metal and carbon have been classified as organometallic, but we shall not discuss these here.

manufacture, from silicon and organic halides was discovered in the 1940s. Again the study of aluminium alkyls has led to their use in catalysts for the large scale polymerization and oligomerization of olefins.

Properties

Most organometallic compounds resemble organic rather than inorganic compounds in their physical properties. Many possess discrete molecular structures, and hence exist at ordinary temperatures as low melting crystals, liquids or gases (Me_3B) (see Table I). Commonly they are soluble in weakly polar organic solvents such as toluene, ethers or alcohols. Their chemical properties vary widely, and, for example, as is true for simple organic compounds their thermal stability depends markedly on their chemical composition. Thus tetramethylsilane (Me_4Si) is unchanged after many days at 500°C, whereas tetramethyltitanium decomposes rapidly at room temperature. Similarly there are wide differences in their kinetic stability to oxidation; some (e.g. Me_4Si, Me_2Hg) are not attacked at room temperature by the oxygen in air, whereas others (e.g. Me_3B, Me_2Zn) are spontaneously inflammable.

Classification of organometallic compounds by bond type

Organometallic compounds may conveniently be classified by the type of metal-carbon bonding which they contain. Carbon is a fairly electro-negative element (2·5 on the Pauling scale), and hence might be expected to form ionic bonds only with the most electropositive elements, but to form electron-pair covalent bonds with other elements. The Periodic Table may be divided into very approximate regions in which the various types of organometallic compound predominantly fall (Figure 1). As is usually found in inorganic chemistry, such a classification is far from rigid, and the regions overlap considerably. It will be noticed that these regions are very similar to those observed in the classification of hydrides into (a) ionic, (b) volatile covalent, (c) covalent, electron-deficient and (d) metal-like or 'interstitial'. As we shall see later, the organic compounds of the d-block transition elements often involve not only σ- but also π- or δ-bonding which is not commonly found amongst compounds of the main group elements. In detailed discussion of their chemistry, therefore, d-block transition metal organometallic compounds are better taken separately from those of the main group elements. Moreover, the chemistry of organic derivatives of the transition elements may be dominated more by the ligand (especially when it occupies in effect several co-ordination positions) rather than by the Periodic Group.

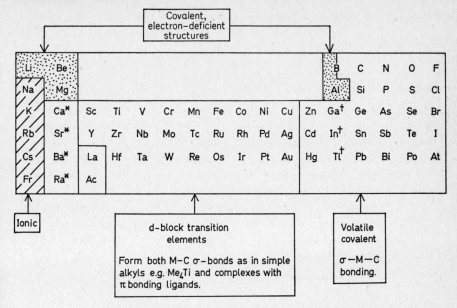

* The nature of the M—C bonding is not clearly established.
† See chapter 3, p. 40–41.
Figure 1. Types of organometallic compounds and the Periodic Table.

Covalent, two-centre, two electron bonds

The simplest type of metal-carbon bond consists of an essentially single covalent 2-electron bond M—C. Depending on the electronegativity of the metal M (and to a lesser extent of R) a wide variation of ionic character in such bonds occurs, ranging from the essentially ionic (see pp 33, 42), $(Ph_3C^-Na^+; (C_5H_5^-)_2Mg^{2+})$ to predominantly covalent bonds such as B—C in Me_3B or Si—C in Me_4Si. For example, the B—C bond in Me_3B (electronegativity difference $x_C - x_B = 2.5 - 2.1 = 0.4$) is less ionic in character than the Al—C bond in the monomeric form of Me_3Al (electronegativity difference $x_C - x_{Al} = 2.5 - 1.6 = 0.9$).

Bond energies

The M—C single bond energies in various compounds are collected in Table I and Figure 2.

(*i*) The metal-carbon single bond occurs widely throughout the Periodic Table. It is not restricted to main group elements, being common among *d*-block transition elements as well, although in the latter case the compounds are usually subject to rather special rules if thermal stability is to be achieved (see pp 7–12, 150–157).

Table I. *Mean metal-carbon bond dissociation energies* $\bar{D}$ *(M—Me)†* *(kcal/mole) and boiling points* (°C, *in parentheses*) *for compounds* Me_nM.

Me_2M	$\bar{D}$, (bp)	Me_3M	$\bar{D}$, (bp)	Me_4M^*	$\bar{D}$, (bp)	Me_3M	$\bar{D}$, (bp)
Me_2Be	−(217)††	Me_3B	87(−22)	Me_4C	83(10)	Me_3N	75(3)
Me_2Mg	—	Me_3Al	66(126)	Me_4Si	70(27)	Me_3P	66(40)
Me_2Zn	42(44)	Me_3Ga	59(56)	Me_4Ge	59(43)	Me_3As	55(52)
Me_2Cd	33(106)	Me_3In	41(136)	Me_4Sn	52(77)	Me_3Sb	52(79)
Me_2Hg	29(93)	Me_3Tl	−(147)**	Me_4Pb	37(110)	Me_3Bi	34(110)

†$n\,\bar{D}$ (M—Me) $= \Delta H° = \Delta H°_f$ (M,g)$+n\Delta H°_f$ (Me,g) $- \Delta H°_f$ (Me$_n$M,g).

*M—C bond lengths for Me$_4$M are, where M = C, 1·54A; Si, 1·87A; Ge, 1·94A; Sn, 2·14A; Pb, 2·30A.

††extrapolated sublimation temp.   **extrapolated bp.

$\bar{D}$ (Pt–Ph) for (Et$_3$P)$_2$PtPh$_2$ = *ca.* 60 kcal/mole.

The dissociation energies quoted are generally ± 1 or 2 kcal/mole.

(*ii*) The bond energies for M—C bonds in the methyl derivatives of some main group elements are plotted against atomic number in Figure 2. This graph shows that within any one main group of the Periodic Table, the metal-carbon bond energy falls with increase in atomic number. Such behaviour is typical of bonds formed between a small first row element such as carbon, and a series of metals in any Group. Overlap between the compact $2s/2p$ hybrid orbitals of carbon (sp^3, sp^2 or sp) is greatest with the valence orbitals of an element in the first Short Period, which are also of principal quantum number 2, than with the larger, more diffuse valence orbitals of higher principal quantum number found in elements of later Periods.

Therefore, while the strengths of single metal-carbon bonds vary from strong (e.g. B—C, Pt—C) to rather weak (e.g. Hg—C, Pb—C), they are generally of a similar order of magnitude to the strengths of single bonds C—X (X = C, N, O, Cl, S, etc.) which are present in familiar organic compounds.

Ionic organometallic compounds

As would be expected, organometallic compounds containing metal ions generally are formed by the most electropositive elements. The formation of ionic compounds is especially favoured when the hydrocarbon anion may be stabilized, for example where the negative charge may be delocalized

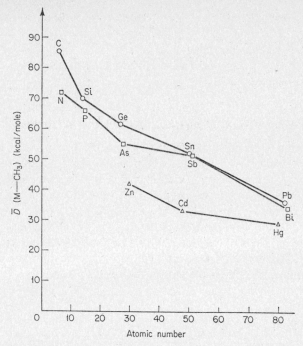

Figure 2. Some mean bond dissociation energies $\bar{D}(M—CH_3)$ in $M(CH_3)_n$.

over several carbon atoms in an aromatic or unsaturated system. The $C_5H_5^{\cdot}$ radical readily accepts an electron giving rise to a $C_5H_5^-$ anion, in which six π-electrons form a delocalized, aromatic system, such as is present in benzene (see p 215). These stable anions (like halide anions) can form salts with cations of electropositive metals which are essentially ionic in character, e.g. $Mg^{2+}(C_5H_5^-)_2$. In sodium acetylide, $(Na^+\bar{C}{\equiv}CH)$, the negative charge is stabilized mainly on account of the greater electronegativity of sp relative to sp^3 hybridized carbon atoms.

Electron-deficient (multicentre bonded) compounds †

In addition to forming either single M—C bonds or salt-like ionic compounds, elements such as Li, Be, Mg, Al, form compounds with bridging alkyl and aryl ligands. This class of compound, which is already familiar

†The term 'electron-deficient' has been criticized on the ground that, if a molecular orbital description of the bonding is used, all bonding m.o.s. which result from combination of available atomic orbitals of suitable energy are, in many cases, filled. The compounds are 'electron-deficient' only in terms of a classical, 2-electron, 2-centre bond picture. The type of bonding postulated here (e.g. in Me_4Li_4 or in carboranes) and which involves multicentre m.o.s. is in some ways similar to that suggested for cluster compounds such as $Mo_6Cl_8^{4+}$ and $Fe_5C(CO)_{15}$.

B

in the boron hydrides, occurs in compounds such as Me_4Li_4, $(Me_2 Be)_n$ and $(Me_3Al)_2$. The occurrence of such compounds may be associated with metals with less than half-filled valency shells, and which would form strongly polarizing cations, i.e. cations having a high charge/radius ratio, e.g. Be^{2+}, (see Chapter 3).

Occurrence of bond type in organo-transition metal complexes

Although d-block transition elements, like main group elements, are capable of forming M—C σ-bonds which are similar in length to bonds between carbon and main group elements of similar atomic number, they have the important additional property of being able to complex with unsaturated hydrocarbons. Here M—C bonds arise essentially from covalent interaction between the π-electron system of the unsaturated hydrocarbon and the metal orbitals. This ability of transition elements to π-bond (see Chapter 5) is associated with electrons in the d-orbitals of the metal, which are of suitable energy for covalent bonding.

Table II. *Summary of Bond Types in Organometallic Compounds*

Ionic	*Covalent, electron-deficient compounds*
The most electropositive elements. Favoured where organic anion is stable.	Multicentre bonds involving carbon. Carbon atom associated with two or more metal atoms in multicentre M.O.s. Electropositive elements where cations would be strongly polarising. Some transition metal compounds, e.g. $Fe_5C(CO)_{15}$ (see p. 181).
σ-Bonding	*π-Bonding*
Essentially covalent two-centre M—C bonds occur with all except the most electropositive Group IA and IIA elements. Both main group and d-block transition elements.	Of covalent character with unsaturated organic ligands. Characteristic of d-block transition elements.

Availability of electron orbitals in metals and metalloids

Orbitals may be divided into three groups as shown in Figure 3 for C, Si (cf. Ge, Sn, Pb) and Ti (cf. Zr, Hf).

Figure 3*.

	C	Si	Ti
Outer Orbitals; empty orbitals which make little or no contribution to bonding, being too high in energy	3d —— 3p —— 3s ——	4p —— 4s ——⎞ 3d ——⎠	5d —— 5p —— 5s ——
Valence Orbitals, which may or may not be fully occupied and which are important in bonding	2p ↿⇂ ↿ ↿ 2s ↿⇂	3p ↿⇂ ↿ ↿ 3s ↿⇂	4p ↿⇂ ↿⇂ ↿⇂ 4s ↿⇂ 3d ↿⇂ ↿⇂ ↿⇂ ↿⇂ ↿⇂
Inner (Core) Orbitals, which are filled and are too low in energy to contribute significantly to the bonding	1s ↿⇂	2p ↿⇂ ↿⇂ ↿⇂ 2s ↿⇂ 1s ↿⇂	3p ↿⇂ ↿⇂ ↿⇂ 3s ↿⇂ 2p ↿⇂ ↿⇂ ↿⇂ 2s ↿⇂ 1s ↿⇂

*The reasons for the variation in the absolute and relative energies of the orbitals throughout the Periodic Table are discussed in detail in Phillips and Williams and in Cotton and Wilkinson (see bibliography, Chapter 1).

Thus although all elements possess *d*-orbitals, only in the case of *d*-block transition elements do they have both suitable energy for bond formation *and* contain electrons.

The 'stability' of organometallic compounds

Introduction

When discussing the stability of a compound one must be quite clear with what type of stability one is concerned. Loose description of a compound as 'stable' may refer to thermal stability, or to resistance to chemical attack, especially to oxidation or hydrolysis. All these aspects of stability depend, as discussed below, on both thermodynamic and kinetic factors.

Thermal stability

In Table III and Figure 4 the heats of formation of a number of main group organometallic compounds are presented. The heat (or strictly the free energy*) of formation of a compound gives a measure of its thermo-

*Few standard entropy data for organometallic compounds are available; consequently $\Delta H_f°$ rather than $\Delta G_f°$ values are used to indicate thermodynamic stabilities. The omission of the entropy term must be borne in mind when considering the data.

Table III. *Heats of formation of some main group organometallic derivatives*

Compound*	ΔH_f° (kcal/mole)	Compound	ΔH_f° (kcal/mole)
EtLi (g)	$13 \cdot 9 \pm 1 \cdot 4$	Me$_4$Si (g)	ca. -57
MeMgI (ether)	$-68 \cdot 3 \pm 0 \cdot 6$	Me$_4$Sn (g)	$-4 \cdot 6 \pm 0 \cdot 6$
Me$_2$Zn (g)	$13 \cdot 1 \pm 2$	Me$_4$Pb (g)	$32 \cdot 6 \pm 0 \cdot 3$
Me$_2$Cd (g)	$26 \cdot 2 \pm 0 \cdot 3$	Me$_3$P (g)	$-23 \cdot 0 \pm 1 \cdot 5$
Me$_2$Hg (g)	$22 \cdot 3 \pm 1$	Me$_3$As (g)	$3 \cdot 7 \pm 1 \cdot 2$
Me$_3$B (g)	$-29 \cdot 3 \pm 5 \cdot 5$	Me$_3$Sb (g)	$7 \cdot 4 \pm 3 \cdot 2$
Me$_3$Al (g)	$-21 \cdot 0 \pm 2$	Me$_3$Bi (g)	$46 \cdot 1 \pm 2$
Me$_6$Al$_2$ (g)	-62 ± 4	Ph$_3$P (c)	$53 \cdot 8 \pm 2 \cdot 5$
Me$_3$Ga (g)	$-9 \cdot 2 \pm 4$	Ph$_3$Bi (c)	$112 \cdot 3 \pm 2$

*(g) indicates gaseous state; (c) indicates crystalline state.

dynamic stability. The low heats of formation of the methyls of the first short period elements (notably Me$_4$C and Me$_3$N) are largely a consequence of the high binding energies of the elements in their standard states (298°K and 1 atm). These data confirm that whereas some organometallic compounds are thermodynamically stable at room temperature (e.g. Me$_4$Si, Me$_3$B) with respect to decomposition to their constituent elements,

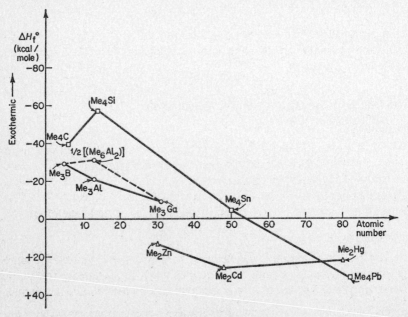

Figure 4. Heats of formation of gaseous methyl derivatives of some main group elements.

others, notably those of the B-elements of the 3rd Long Period, (viz. Me_2Hg, Me_3Tl, Me_4Pb) are unstable to such decomposition, i.e. they are endothermic compounds.

All such endothermic compounds and many more are thermodynamically unstable to reactions such as:

$$R_nM \longrightarrow \text{Hydrocarbons} + \text{metal}$$

an example is

$$Me_2Hg \longrightarrow CH_4 + C_2H_6 + C_2H_4 + Hg$$

Why is it, then, that many of these compounds are isolable, and in some cases show considerable resistance to thermal decomposition? The reason for this is that in addition to a favourable free energy change for the decomposition, a reaction path of sufficiently low energy must also be available for it to occur at a measurable *rate*. In other words, the decomposition of isolable yet thermodynamically unstable compounds may be *kinetically* controlled. This is illustrated in Figure 5. In order for decomposition to proceed, the activation energy or energies for a multi-stage process (strictly free energy of activation) must be sufficiently low. Low activation energies imply a low energy pathway for the rate-controlling step in the decomposition. Such a step may in many cases involve breaking

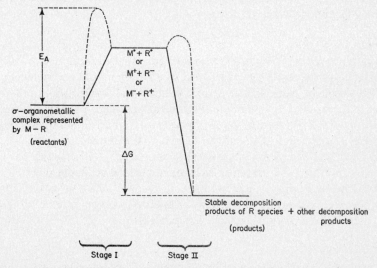

Figure 5. Schematic representation of the thermal decomposition of a σ-organo complex, M—R, by unimolecular dissociation of the M—R bond, showing E_A, the energy of activation of the preliminary decomposition, which is important in determining the kinetic stability of the complex; ΔG represents the free energy of decomposition. If the products have less energy than the reactants then the complex is thermodynamically unstable.

the M—C bond to form M˙ and R˙ radicals (homolytic dissociation) or M$^+$ and R$^-$ or M$^-$ and R$^+$ ions (heterolytic dissociation). Bond dissociation is likely to be facilitated where empty low-lying orbitals are present in the metal atom. In some compounds the activation energy will depend qualitatively on the strength of the metal-carbon bond to an extent related to the degree of bond breaking in the transition state. For example, in Group IVB methyls, as the strengths of the M—C bonds decrease down the group, thermal decomposition of Me$_4$Pb will be favoured compared with Me$_4$C on both kinetic and thermodynamic grounds.

In both homolytic and heterolytic processes, the carbon fragments would normally be very reactive and readily form stable products, for example by dimerization or polymerization. It is the formation of these more stable products which provides energy to drive the decomposition and, since the products would usually be kinetically unreactive, the decomposition would be irreversible under normal conditions. In contrast, dissociation of M—X giving stable neutral ligands such as CO or ethylene, or heterolytic fission forming stable ions such as C$_5$H$_5$$^-$ or Cl$^-$, might be expected to be readily reversible, in which case subsequent decomposition of the dissociated ligand would not normally occur. Therefore in order to be thermally stable, covalent metal-σ-organo complexes must not readily dissociate into reactive fragments, i.e. there should be a high-activation free energy barrier towards dissociation.

Stability to oxidation

All organometallic compounds are thermodynamically unstable to oxidation, the driving force being provided by the large negative free energies of formation of metal oxide, carbon dioxide and water. Many are also kinetically unstable to oxidation at room temperature, for example nearly all the methyls of the main group elements are rapidly attacked, although Me$_2$Hg and the derivatives of the Group IVB elements are inert. Many (e.g. Me$_2$Zn, Me$_3$In and Me$_3$Sb) are spontaneously inflammable in air. Kinetic instability to oxidation may be associated with the presence of empty low-lying orbitals, e.g. $5p$ in Me$_3$In, or of a 'lone pair' of electrons, e.g. Me$_3$Sb. In contrast, the Group IVB alkyls possess neither of these features, and behave as saturated compounds.

Stability to hydrolysis

Hydrolysis of an organometallic compound often involves nucleophilic attack by water, and hence is facilitated by the presence of empty low-lying orbitals on the metal atom. In agreement with this, the organic derivatives of the elements of Groups IA and IIA, and of Zn, Cd, Al, Ga and In are readily hydrolysed. The *rate* of hydrolysis is dependent on the polarity of the M—C bond; where this is high (e.g. Me_3Al), rapid attack by water occurs, whereas Me_3B is unaffected by water at room temperature, even though an empty $2p$ orbital is present on the boron atom. The alkyls and aryls of Group IVB and VB elements, however, are kinetically stable to hydrolysis by water. In these compounds the metal atom is surrounded by a filled shell of 8 electrons, so that nucleophilic attack is no longer favoured. The majority of the neutral organic derivatives of transition metals are inert to hydrolysis.

General features relating to stability: filled shells of electrons

It is well known that much of carbon chemistry is controlled by kinetic factors; for example diamond would change spontaneously to graphite, and acetylene to benzene, at room temperature and atmospheric pressure if thermodynamics were controlling. In addition, nearly all organic compounds are thermodynamically unstable to oxidation, and exist in the presence of air only because no suitable low energy oxidation mechanism is available.

This 'kinetic stability' of carbon compounds has a variety of causes, notably the *full* use of the four valence orbitals (sp^3) in carbon (leading to the common maximum co-ordination number of four—exceptions, e.g. Me_4Li_4, $(Me_3Al)_2$ in which the co-ordination number rises to 5–7 are discussed in Chapter 3), and the high energy of empty antibonding or non-bonding orbitals into which electrons could either be promoted to initiate thermal decomposition or donated in the case of nucleophilic attack.

In silicon the $3d$ orbitals lie relatively close in energy to the $3s$ and $3p$ valence electrons (energy gap 129 kcal/mole, see Figure 3) and may in certain cases contribute significantly to the bonding, e.g. in $(H_3Si)_3N$ or $SiF_6^=$, or in a transition state in their reactions, although the extent of their use is difficult to assess. Expansion of the co-ordination number above four commonly occurs in compounds of silicon and of other main group elements of the 2nd and later periods, when strongly electronegative groups (e.g. halogen atoms) are attached to the metal atom (e.g. $SiF_6^=$). The occurrence of co-ordination numbers above four and the kinetic instability of $SiCl_4$ to hydrolysis, in contrast to CCl_4, may be associated with the presence of empty $3d$ (also $4s$, $4p$) orbitals in the former which can accept electrons from water in nucleophilic attack. As we have seen, however, where such electron-attracting groups are absent (e.g. Me_4Si) there is generally no tendency for the covalency of the metal atom to expand above 4, so that Me_4Si is *kinetically* stable to thermal decomposition, oxidation and hydrolysis at room temperature. *Thus kinetic stability of organometallic compounds may be associated with a closed shell of electrons, often of essentially spherical symmetry, around the metal atom.*

For compounds of the transition elements, however, empty valence shell (e.g. $3d$, $4d$ or $5d$) orbitals are often available, and can markedly decrease their kinetic stability. For example Me_4Ti is unstable at room temperature, decomposing to hydrocarbons and other products, whereas Me_4Si may be heated to at least 500°C. In transition metals the closed shell concept remains but can apply in several different ways. The commonest of these is where the closed shell consists of 18 electrons, i.e. ns^2, np^6, and $(n-1)d^{10}$. Transition metal atoms are, like atoms of Group IA and IIA elements, inherently electron-deficient, and this electron-deficiency must be satisfied if thermally stable organometallic compounds are to be formed. The 18-electron rule and related ideas are discussed further in Chapter 5.

BIBLIOGRAPHY

Suitable background reading

F. A. Cotton and G. Wilkinson, 'Advanced Inorganic Chemistry', 2nd Edition (Interscience, New York, 1966).

C. S. G. Phillips and R. J. P. Williams, 'Inorganic Chemistry', Volumes I and II (Oxford University Press, Oxford, 1966).

J. D. Roberts and M. C. Caserio, 'Basic Principles of Organic Chemistry' (Benjamin, New York, 1965).

Previous general texts on organometallic chemistry

G. E. Coates, 'Organometallic Compounds', 2nd Edition (Methuen, London, 1960) A useful but slightly out of date account.

J. J. Eisch, 'The Chemistry of Organometallic Compounds, the main group elements', (MacMillan, New York, 1967).

P. L. Pauson, 'Organometallic Chemistry' (Edward Arnold, London, 1967).

More advanced texts on organometallic chemistry

G. E. Coates, M. L. H. Green and K. Wade, 'Organometallic Compounds', 3rd Edition. Volume I 1967, 'The main group elements'; Volume II 1968, 'The transition elements' (Methuen, London).

E. Krause and A. von Grosse, 'Die Chemie der metallorganischen Verbindungen', (Borntraeger, Berlin, 1937). An excellent review, in German, of early work.

'Metal Organic Compounds', *Advances in Chemistry Series*, No 23 (American Chemical Society, Washington, D.C., 1959). A collection of papers on various aspects.

Annual reviews of research work

Annual Reports (Chemical Society, London). Reviews of recent research, with many references, including sections on organometallic compounds, metal carbonyls, π-complexes, etc.

Annual Surveys of Organometallic Chemistry, D. Seyferth and R. B. King (Elsevier, Amsterdam, 1964–66). Useful, comprehensive summaries of recent research.

Thermochemistry of organometallic compounds

C. T. Mortimer, 'Reaction Heats and Bond Strengths', (Pergamon, London, 1962). A useful textbook, which contains a chapter on the thermochemistry of organometallic compounds.

H. A. Skinner, *Advances in Organometallic Chemistry* (Ed. F. G. A. Stone and R. West Academic Press, New York), 1964, **2**, 49. An excellent review including many data.

CHAPTER TWO

Methods of formation of metal-carbon bonds of the main group elements

The reaction between a metal and an organic halogen compound

$$2M + n\,RX = R_nM + MX_n \text{ (or } 2\,R_nMX_n)$$

The reaction between a metal and an organic halogen compound provides an important and fairly general route to organometallic compounds. It is interesting to note that one of the earliest organometallic compounds ever to be prepared, ethylzinc iodide, was made in this way from zinc/copper couple and ethyl iodide.

Thermochemical considerations

It is useful to examine the scope of the reactions between metals and organic halides in the light of their thermochemistry. Some typical values of the heats of reaction between methyl halides and main group elements are given below (Figures 6, 7). Heats rather than free energies of reaction are used, as negligible data on the standard entropies of organometallic compounds are available. Within one series of reactions involving compounds of the same group of the Periodic Table, however, where the stoicheiometry is the same for each element, the neglect of the entropy term should not, to a first approximation, seriously alter the gross pattern of changes within the Group.

These data show that, apart from the elements C, N and O and the B-elements of the last row of the Periodic Table (Hg, Tl, Pb and Bi), the reaction between an element and methyl chloride to yield the chloride and methyl of the element is exothermic, and therefore might be expected to provide a method of synthesizing the methyls. In all the examples listed where a negative value of ΔH is found, such reactions have been observed, and many are important synthetic routes. The mechanisms of the reactions and the conditions under which they occur, however, vary greatly with different elements. Kinetic as well as thermodynamic factors are thus important in controlling the reactions.

Figures 6 and 7 show that the main driving force for these reactions is, in general, the strongly exothermic formation of the metal halide. This is particularly striking for the Group IIB elements Zn, Cd and Hg, where the methyls are in each case endothermic compounds. Although Cd and Hg react directly with methyl iodide only in the presence of light, which initiates homolytic dissociation of the organic halide into $\cdot CH_3$ and $I\cdot$, zinc as zinc/copper couple reacts readily to yield methylzinc iodide.

In the B-elements the strength of the M—C bonds falls sharply on going from the elements of the second long Period (e.g. Sn, Sb) to the elements of the third long Period (e.g. Pb, Bi). The alkyls of Hg, Tl, Pb and Bi are all strongly endothermic compounds, and their unfavourable heats of formation are not outweighed by the formation of the exothermic halides when the elements are treated with alkyl halides.

Calculated heats of reaction of methyl halides with metals suggest that CH_3F should react more exothermically than the other halides. In practice, however, methyl fluoride reacts by far the most reluctantly. Although the detailed mechanism of attack of a methyl halide on a metal will vary with the metal concerned, the activation energy will almost certainly be related

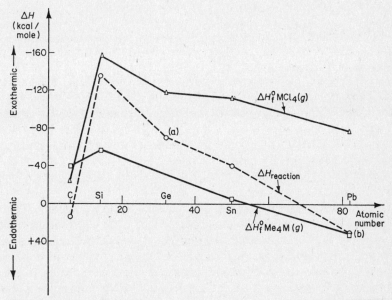

Figure 6. Thermochemistry of reactions $2M + 4CH_3Cl(g) \rightarrow MCl_4(g) + M(CH_3)_4(g)^{(c)}$.
  (a)   estimated.
  (b)   Point refers to $PbCl_4(l)$.
  (c)   In fact mixed methyl chlorides Me_nMCl_{4-n} are the thermodynamically favoured products.

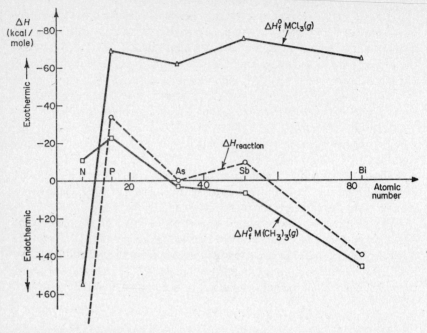

Figure 7. Thermochemistry of the reactions $2M + 3CH_3Cl(g) \rightarrow MCl_3(g) + M(CH_3)_3(g)^{(a)}$.
(a) In fact mixed methyl chlorides Me_nMCl_{3-n} are the thermodynamically favoured products.

to the strength of the carbon-halogen bond which is breaking. The C—Hal bond energies fall from C—F (ca. 105), C—Cl (ca. 79), C—Br (ca. 66) to C—I (ca. 57 kcal/mole), whereas the relative reactivities (rates) of organic halides to metals usually rise: $RF \ll RCl < BRr < RI$. This is another example of reactions which are thermodynamically very favoured, but which are controlled by kinetic factors.

Applications

The reaction between metals and organic halides is often suitable for the synthesis of organometallic compounds of the most electropositive elements. On a laboratory scale, the derivatives of Li, Mg and Al, which are very important in synthetic work, are usually prepared in this way. Some illustrations of the method are given briefly here; more detailed discussion of the preparation and properties of the compounds involved appears in Chapter 3.

Organolithium compounds may be prepared by the action of lithium metal on alkyl or aryl halides, in ether or hydrocarbon solvents,

$$2Li + RX = RLi + LiX$$

Chlorides are preferred to bromides in the alkyl series, and with the important exception of methyl iodide, iodides cannot be used since they react too rapidly with the organolithium compound,

$$RLi + RI = R{-}R + LiI \text{ (Wurtz Coupling).}$$

This coupling reaction is less serious in the aryl series, and bromides or iodides can often be used successfully.

Magnesium reacts with alkyl and aryl halides in ethers to form Grignard reagents:

$$RX + Mg = RMgX$$

For R = alkyl, chlorides are used whenever possible. They are cheaper than the bromides and iodides, and although they react more slowly with magnesium, side reactions such as Wurtz coupling or olefin elimination are not so serious. Yields of Grignard reagents from primary alkyl chlorides or bromides are usually better than 80%. Secondary and tertiary halides, especially bromides and iodides, are not so satisfactory. With tert-butyl bromide, for example, much butene is evolved by elimination, and only low yields of tert-butylmagnesium bromide are obtained. Alkyl fluorides react too slowly to be convenient materials from which to prepare Grignard reagents RMgF, on account of the kinetic inertness of the C—F bond to attack; good yields of alkylmagnesium fluoride are formed only during several days in refluxing tetrahydrofuran with e.g. iodine or ethylene dibromide as catalyst.

Aluminium reacts with a limited number of aryl and alkyl halides, forming $R_3Al_2X_3$. These reactions are further considered in Chapter 3.

When volatile organic halides (the chlorides are almost always used) are passed over silicon, mixed with a metal such as copper and heated at 250–400°C, organosilicon compounds, notably R_2SiCl_2, are produced,

$$Si + 2 \, RCl = R_2SiCl_2 \text{ (R = alkyl or aryl)}$$

The organosilicon halides are important intermediates in silicone manufacture. This reaction is often called the 'Direct' synthesis, and with its applications, is discussed in Chapter 4. It is not restricted to silicon chemistry, however, although it is in this field that its main importance lies. Similar reactions occur between alkyl halides and metals and metalloids such as Ge, Sn, P, As and Sb. As long ago as 1870 Cahours observed that arsenic reacts with methyl iodide in a sealed tube,

$$4MeI + 2As = Me_4\overset{+}{As}\,\overset{-}{I} + AsI_3$$

More recently (1957), it has been shown that when methyl bromide is passed over a heated arsenic/copper mixture the following reaction occurs

$$3MeBr + 2As = MeAsBr_2 + Me_2AsBr$$

Alloy methods

So far we have considered only the reactions of the more electropositive elements. We have seen that for the heaviest elements (e.g. Hg, Tl, Pb, Bi) the alkyls and aryls are endothermic compounds and that even the reactions between the element and organic halide may be endothermic, in spite of a high heat of formation of a metal halide tending to drive the reaction to the right. In these cases the direct synthesis can still be used by taking an alloy of the metal with an alkali metal such as sodium, instead of the free metal. Here the formation of the strongly exothermic sodium halide provides the driving force. In this way alkyls and aryls of Hg, Tl, Sn, Pb and Bi can be made, e.g.

$$2Hg(l) + 2MeBr(l) = Me_2Hg(l) + HgBr_2(s) \quad \Delta H = 0 \text{ kcal/mole}$$
$$HgBr_2(s) + 2Na(s) = Hg(l) + 2NaBr(s) \quad \Delta H = -132 \text{ kcal/mole}$$

$$Hg(l) + 2MeBr(l) + 2Na(s) = Me_2Hg(l) + 2\,NaBr(s) \quad \Delta H = -132 \text{ kcal/mole}$$

The alloy method is used in the manufacture of tetramethyl- and tetraethyl-lead, anti-knock additives to petrol. These are the largest scale industrial organometallic products.

$$4EtCl + 4NaPb = Et_4Pb + 3Pb + 4NaCl$$

Similarly bismuth triaryls were first made by Michaelis in 1877 by the reaction of powdered bismuth-sodium alloy with aryl halides,

$$3ArX + Na_3Bi = Ar_3Bi + 3NaX$$

Metal exchange: the reaction between a metal and an organometallic compound of another metal

$$M + RM' = M' + RM$$

This method depends on the difference in the free energies of formation of the two species RM' and RM. As $\Delta G°$ values for organometallic compounds are available in very few cases, $\Delta H°$ (heats of formation) will be used to indicate the thermodynamic feasibility of such reactions.

It is expected that endothermic or weakly exothermic organometallic compounds will be the most versatile reagents RM' in such reactions. Such compounds are found among the heavy B elements (Hg, Tl, Pb and Bi). Of these the air-stable and readily prepared (though toxic) organo-mercury compounds are the most widely used. Let us consider the heats of reaction between dimethyl mercury and zinc and cadmium:

(i) $Zn(s) + Me_2Hg(l) = Me_2Zn(l) + Hg(l) \quad \Delta H = -8 \text{ kcal/mole}$

(ii) $Cd(s) + Me_2Hg(l) = Me_2Cd(l) + Hg(l) \quad \Delta H = +2 \cdot 7 \text{ kcal/mole}$

This shows that the zinc reaction goes essentially to completion whilst an equilibrium mixture is observed in the cadmium system. It is therefore inadvisable to handle volatile organocadmium compounds in vacuum apparatus containing mercury, as the reverse of reaction (ii) can occur.

Other elements which react satisfactorily with dialkyl and diaryl mercury compounds include the alkali metals, the alkaline earths, Al, Ga, Sn, Pb, Sb, Bi, Se and Te (see Chapters 3, 4). In certain cases the formation of a metal amalgam assists the reaction. The transition elements, when they react, give metallic mercury and a mixture of hydrocarbons. An unstable organotransition metal species may be formed initially, only to decompose. The reactions,

$$\Delta H \text{ (kcal/mole)}$$

$$P(\text{white}) + Ph_3Bi(c) = Bi(s) + Ph_3P(c) \qquad -58$$

$$As(s) \quad + Ph_3Bi(c) = Bi(s) + Ph_3As(c) \qquad -42\cdot2$$

$$Sb(s) \quad + Ph_3Bi(c) = Bi(s) + Ph_3Sb(c) \qquad -33\cdot6$$

are similar. It should be noted that the heat (or more strictly, free energy) of formation of the organoelement compound is the controlling factor. The order of displacement of one element by another in such reactions has in the past often been related to the electrochemical series. An element high in this series will displace another element from its organometallic derivatives just as from its salts. Such a rule holds for the metallic elements. As the above examples show, however, where non-metals or metalloids such as P, As, Sb and Bi are concerned, it is better and more meaningful to consider $\Delta H_f°$ rather than any position in the electrochemical series.

The transfer of alkyl or aryl groups from one element to another may not necessarily be complete. Thus the aryls of Groups IVB and VB react with alkali metals in ether solvents or in liquid ammonia, for example,

$$Ph_4Sn + 2Na \xrightarrow[\text{NH}_3]{\text{liq.}} Ph_3SnNa + NaPh \text{ (attacks solvent)}$$

$$Ph_3P \; + 2Na \xrightarrow{\text{THF}} Ph_2PNa + NaPh \qquad \text{''}$$

Reactions of organometallic compounds with metal halides

$$RM + M'X = RM' + MX$$

These provide another important and general method for the preparation of organo-element compounds; they are probably the most widely used and versatile of all laboratory methods. In order to understand the direction in which any such reaction will go, let us consider the thermochemical data given diagrammatically in Figure 8. Here the differences between the heats of formation of the chlorides and the methyls of the same element,

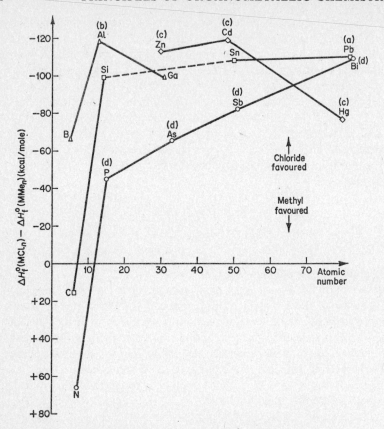

Figure 8. Plot of ΔH_f° [MCl$_n$(g)] $-\Delta H_f^\circ$ [MMe$_n$(g)] against atomic number.
    (a) Using ΔH_f° [PbCl$_4$(l)]   (b) Monomeric Me$_3$Al(g).
    (c) Using ΔH_f° [MCl$_2$(c)].   (d) Trivalent MX$_3$.

ΔH_f° (MCl$_n$)$-\Delta H_f^\circ$ (MMe$_n$), are plotted against atomic number for a number of B elements. It will be seen that as one descends Groups IVB (C to Sn) or Group VB (N to Bi), the chloride becomes increasingly favoured with respect to the methyl. This is also the general trend of increasing metallic or electropositive character of the elements. Thus, generally, the effect is to combine the halogen with the more electropositive element and the organic radical with the more electronegative element of the pair.

By far the most versatile reagents in such syntheses are the organic derivatives of the most electropositive elements, especially, on account of their easy preparation, those of Li and Mg. These react rapidly and exo-thermically with halides of more electronegative elements. Examples are very numerous so that only a few are given here to illustrate the point:

$$2RMgX + CdCl_2 = R_2Cd + 2MgXCl$$

Cadmium alkyls are useful reagents in organic chemistry for the conversion of acyl halides to ketones. They are prepared *in situ* by the direct addition of anhydrous $CdCl_2$ to an ether solution of the Grignard reagent.

Organolithium compounds are more reactive than Grignard reagents, and in certain cases allow more complete substitution. For example, Grignard reagents substitute only two of the chlorine atoms in thallium (III) chloride when diethyl ether is the solvent, whereas lithium alkyls substitute all three:

$$Et_3Tl \xleftarrow[\text{ether}]{\text{EtLi}} TlCl_3 \xrightarrow[\text{ether}]{\text{EtMgCl}} Et_2TlCl$$

Similarly $(Et_3P)_2PtBr_2$ yields the dimethyl derivative $(Et_3P)_2PtMe_2$ with methyl-lithium, whereas the Grignard reagent gives only the monomethyl complex $(Et_3P)_2PtMeBr$. (see p 174).

Another factor which influences the choice between organolithium and Grignard reagents is the hydrolytic stability of the product. Products from a Grignard reaction are most conveniently isolated by hydrolysis of the reaction mixture with dilute hydrochloric acid, saturated ammonium chloride solution or water to destroy any excess Grignard reagent and to dissolve out magnesium salts, followed by separation of the organic layer, drying, and removal of the solvent. Omission of the hydrolysis step leads to a rather intractable mixture of the desired product with sticky ether complexes of magnesium halides. The products are especially difficult to separate when bromides and iodides are present. Lithium alkyl, on the other hand, can be prepared in petroleum ether in which lithium halides are quite insoluble. Water-sensitive products can be isolated by filtration from the lithium halide and evaporation of the filtrate in a dry atmosphere.

Insertion of olefins and acetylenes into metal-hydrogen bonds

These reactions may be represented by the general equation:

$$M-H + \underset{/}{\overset{\backslash}{C}} = \underset{\backslash}{\overset{/}{C}} \rightleftharpoons M-\overset{|}{\underset{|}{C}}-\overset{|}{\underset{|}{C}}-H$$

The most important examples concern B, Al, and Si. (Additions of non-metal hydrides N—H, P—H, O—H, S—H and H—Hal to olefins and acetylenes are not included here). These reactions are not as simple as the equation suggests, and it has been shown that several different mechanisms can operate, even in the reactions of hydrides of one element such as silicon.

c

Boron and aluminium

The addition of boron hydrides and of aluminium hydrides to olefins are processes of wide application. The former, known as hydroboration, provides a simple laboratory route to boron alkyls, and thence to a wide range of organic compounds from readily available starting materials, (Chapter 3, p 71). The latter is important industrially and is discussed in the section on aluminium, (Chapter 3, p 91). Addition of boron hydrides (usually diborane or alkylboron hydrides such as $(RBH_2)_2$ or $(R_2BH)_2$) to olefins occurs under mild conditions at room temperature in an ether solvent. The role of ethers is unclear; they may catalyze some reactions, whilst in other cases they are not essential.

Recently, the corresponding addition of Ga—H bonds to olefins has been described. For example, Et_2GaH adds 1-decene giving $Et_2Ga\,C_{10}H_{21}$, and the dimer $(GaHCl_2)_2$ also adds terminal olefins,

$$(HGaCl_2)_2 + RCH{=}CH_2 \xrightarrow[\text{room temp}]{\text{ether}} (RCH_2CH_2GaCl_2)_2$$

Addition of transition metal hydrides to olefins and the role of such reactions in catalytic processes are discussed in Chapter 9.

Addition of Group IVB hydrides to olefins

As one descends Group IVB of the Periodic Table from Si to Ge to Sn to Pb, the ease of addition of the hydrides (e.g. R_3MH) to olefins increases markedly with the decrease in strength of the M—H bonds. Thus trialkyl-silanes require either heating to about 300°C under olefin pressures of at least 300 atm, or UV irradiation, or catalysis by, for example, peroxides, certain metal salts or tertiary amines. Trialkylgermanes and trialkyl-stannanes add to activated terminal carbon-carbon double bonds at 120°C and 90°C respectively, and a catalyst is not necessary, e.g.

$$Bu_3GeH + H_2C{=}CHCN \longrightarrow Bu_3GeCH_2CH_2CN$$

Tri-*n*-butyl-lead hydride is so reactive that it adds rapidly to olefins and acetylenes in ether solution without a catalyst even at 0°C, e.g.

$$Bu_3PbH + PhC{\equiv}CH \longrightarrow trans\text{-}PhCH{=}CHPbBu_3.$$

Hydrosilation

The addition of Si—H bonds to olefins and acetylenes is of some importance industrially in the production of intermediates for silicone manufacture.

$$X_3SiH + H_2C{=}CHY \longrightarrow X_3SiCH_2CH_2Y$$
$$X_3SiH + HC{\equiv}CH \longrightarrow X_3SiCH{=}CH_2 \longrightarrow X_3SiCH_2CH_2SiX_3$$

Addition may occur by thermal reaction under pressure, but proceeds better under UV irradiation or when catalyzed. The thermal, photolytic or peroxide catalyzed reactions involve free-radical chain mechanisms. Radicals are first produced from the silicon hydride,

$$X_3SiH \xrightarrow{h\nu} X_3Si\cdot + H\cdot$$

which then add to the olefin by various pathways,

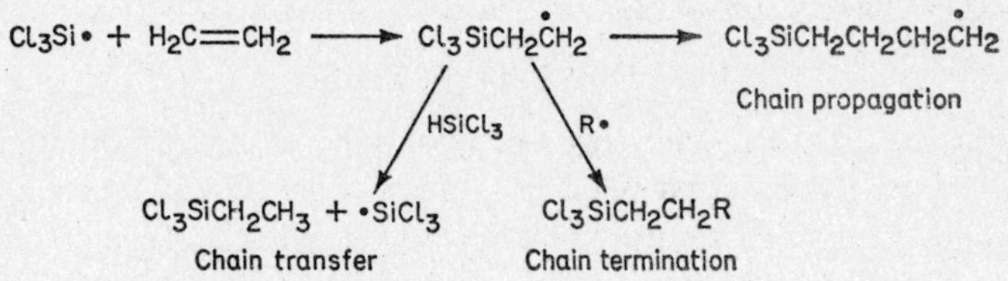

The chain propagation step is often not very important, as the silyl radicals are very efficient chain transfer agents.

Catalysis by transition metal salts, e.g. H_2PtCl_6, $[(\text{cyclohexene})PtCl_2]_2$, probably involves an olefin-complex of Pt(II) and intermediates with covalent Pt—Si bonds.

Formation of metal-carbon bonds by other insertion reactions

The addition of metal hydrides to carbon-carbon double or triple bonds is a special case of a very general reaction type, which involves the addition of a species A—B to an unsaturated system X=Y (or X≡Y).

$$A-B + X=Y \longrightarrow A-X-Y-B$$

In this section we discuss examples of this general type, which involve the addition of bonds between metals and elements other than hydrogen (A = Metal; B = −C, −N, −P, −O, −Hal, −Metal, etc.), to olefins and acetylenes. Formally similar reactions which involve addition of these bonds to species X=Y, where Y ≠ C (e.g. to C=O, C=N, etc.) are described briefly elsewhere, as these do not lead to the formation of new M—C bonds.

Polar metal-carbon bonds, such as are found in the organometallic compounds of the most electropositive elements, can add to olefins and acetylenes, especially when the latter bear electron-withdrawing substituents, which either increase the polarity of the C=C or C≡C bond or can stabilize a negative charge in the transition state. Thus butyl-lithium adds to diphenylacetylene in diethyl ether, thought not in pentane, and *trans*-butylstilbene may be isolated after hydrolysis, showing that *cis*- addition occurs:

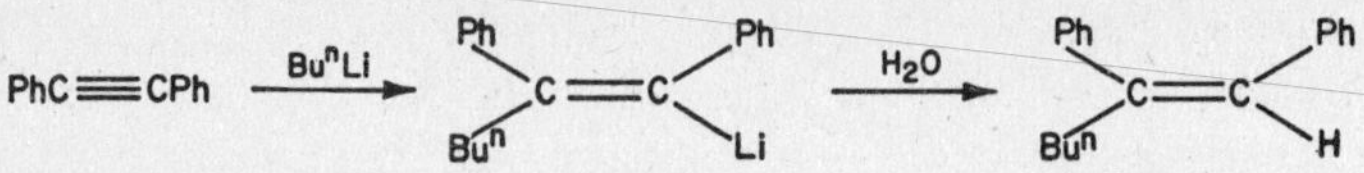

$$PhC\equiv CPh \xrightarrow{\text{Bu}^n\text{Li}} \underset{Bu^n}{\overset{Ph}{\diagup}}C=C\underset{Li}{\overset{Ph}{\diagdown}} \xrightarrow{H_2O} \underset{Bu^n}{\overset{Ph}{\diagup}}C=C\underset{H}{\overset{Ph}{\diagdown}}$$

A series of insertion reactions occurs when n–butyl-lithium reacts with ethylene under high pressure (10 000 to 15 000 psi). Waxy polythene of molecular weight up to about 17 000 is produced,

$$Bu^nLi + H_2C=CH_2 \longrightarrow Bu^nCH_2-CH_2Li \xrightarrow{nH_2C=CH_2}$$
$$Bu^n(CH_2CH_2)_{n+1}Li. \text{ (see p 51)}$$

Telomerisation of ethylene also occurs with triethylaluminium (see Chapter 3, p 93). Grignard reagents, unlike organolithium and organo-aluminium compounds, do not normally add to olefins or acetylenes.

The addition of M—N, M—P, M—O and M—M to olefins and acetylenes sometimes occurs, especially when these bonds are weak, as with the heavier elements such as Sn and Pb. Such reactions are favoured when the olefin or acetylene bears electron-attracting substituents such as –COOMe or –CN. Many of these studies have been carried out with derivatives of tin, and some examples from the chemistry of this element are given below. Most of this work is still in an exploratory stage and applications outside the laboratory have not yet been developed.

$$Et_3Sn-OMe + H_2C=C=O \longrightarrow Et_3Sn-CH_2C\overset{O}{\underset{OMe}{\diagup\diagdown}}$$

$$Ph_3Sn-PPh_2 + PhCH=CH_2 \longrightarrow Ph_3SnCH_2CH\underset{PPh_2}{-}Ph \text{ (catalyzed by free radical sources).}$$

$$Me_3Sn-SnMe_3 + C_2F_4 \longrightarrow Me_3SnCF_2CF_2SnMe_3$$

The related mercuration of olefins is discussed on p 28.

Reactions of diazo compounds

The use of diazo compounds in the formation of metal-carbon bonds is conveniently divided into two sections—the reactions of (a) aliphatic and (b) aromatic diazo compounds.

(a) Aliphatic diazo compounds

Diazomethane and substituted diazomethanes (e.g. diazoacetic ester) react with many metal halides or metal hydrides under mild conditions by methylene insertion:—

$$SiCl_4 + CH_2N_2 \xrightarrow[\text{ether}]{-50°C} Cl_3SiCH_2Cl + N_2$$

$$Me_2SnCl_2 + CH_2N_2 \xrightarrow{\text{ether}} Me_2Sn(Cl)CH_2Cl + N_2$$

$$HgCl_2 + CH_2N_2 \longrightarrow ClCH_2HgCl \xrightarrow{CH_2N_2} ClCH_2HgCH_2Cl$$

$$R_3SnH + N_2CHCOOEt \longrightarrow R_3SnCH_2COOEt + N_2$$

Replacement of all the M—Cl groups is often difficult or impossible.

Where the reactant is a strong electron-pair acceptor (Lewis acid) e.g. X_3B (X = Cl, R) or X_3Al, the formation of polymethylene is a serious complication. In these cases the initial step involves co-ordination of the CH_2 group of diazomethane with the vacant p–orbital on boron,

$$X_3B + H_2\bar{C}\text{—}\overset{+}{N}\text{≡≡}N \longrightarrow X_3\bar{B}\text{—}CH_2\text{—}\overset{+}{N}\text{≡≡}N$$

this is followed in fast subsequent steps by elimination of nitrogen and polymerization:

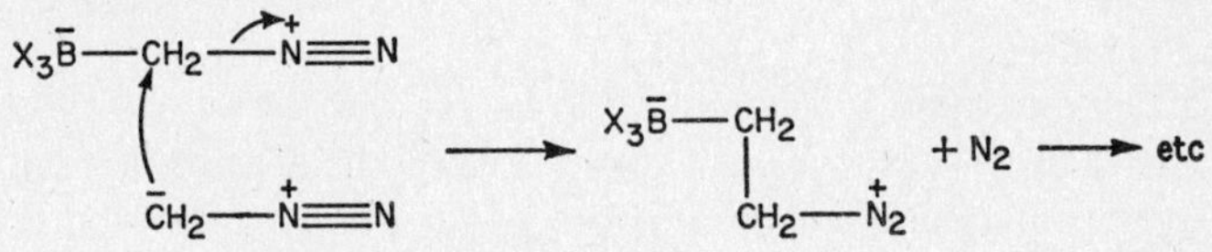

(b) *Reaction of aromatic diazonium salts with metal and metalloid halides or oxides in aqueous solution*

This method is applicable to a wide range of heavy metal and metalloid elements. Two conditions must be fulfilled for it to be applied successfully. First, as aqueous media are used, the product must not be susceptible to hydrolysis. This restricts its use to aryls of Hg, Tl and the Group IVB and VB elements. Secondly the element involved must be susceptible to electrophilic attack by the aryldiazonium ion ArN_2^+.

The best known, and probably the most successful applications are the Bart reaction for the preparation of arylarsonic acids, and the analogous Schmidt reaction in antimony chemistry, e.g.

$$PhN_2X + As(OH)_3 \xrightarrow[\text{aq. acetone}]{\text{alkaline}} PhAsO(OH)_2 + N_2 + HX$$

$$PhN_2X + Sb(OH)_3 \xrightarrow[\text{aq. acetone}]{\text{alkaline}} PhSbO(OH)_2 + N_2 + HX$$

It should be noted that oxidation of the As or Sb from the trivalent to the pentavalent state occurs.

Nesmeyanov has extended this reaction to the synthesis of aryl deriva-tives of the heavier B–elements, e.g. of Hg, Tl, Sn, Pb and Bi, but the yields are often low. Often the double salts of the metal halide with the diazonium salt are first isolated, and then reduced by a metal such as copper or zinc, e.g.

$$[PhN_2]^+[HgI_3]^- + 2Cu \longrightarrow PhHgI + N_2 + 2CuI$$

$$2PhN_2Cl + SnCl_4 \longrightarrow [PhN_2]_2{}^+[SnCl_6]^-$$

$$\downarrow 2Zn$$

$$Ph_2SnCl_2 + 2N_2 + 2ZnCl_2$$

Decarboxylation of heavy B-metal salts

As is well known, the decarboxylation of the calcium or barium salts of aliphatic acids leads to ketones, e.g.

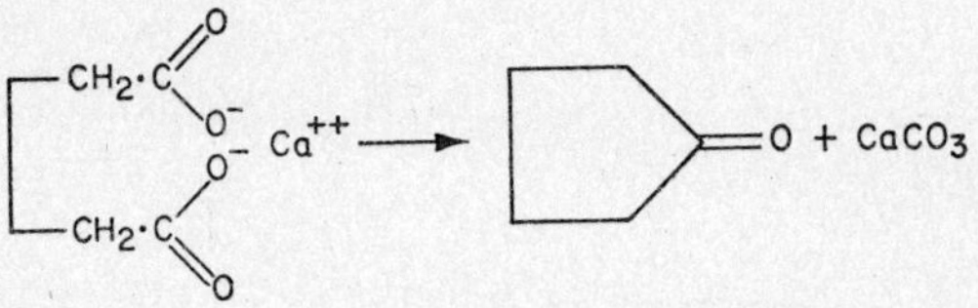

The decarboxylation of certain salts of organic acids with heavy B–elements, however, leads to the formation of metal-carbon bonds. It is generally necessary for the organic acid to contain electron-withdrawing groups. Most examples come from Hg, Sn or Pb chemistry, e.g.

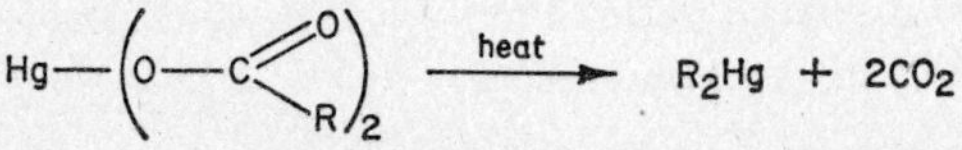

$$(R = C_6F_5; C_6Cl_5; CF_3; 2,4,6\text{-trinitrophenyl}; CCl_3)$$

Contrast the effect of heat on mercuric benzoate, when substitution of the aromatic ring *ortho* to the carboxyl group occurs. Mercuric tri-chloroacetate decarboxylates so rapidly that it has not yet been isolated. Bis(trichloromethyl)mercury was obtained in good yield by the reaction of sodium trichloroacetate and mercuric halides in ethyleneglycol-dimethyl ether:

$$2Cl_3C.CO_2Na + HgCl_2 = (Cl_3C)_2Hg + 2NaCl + 2CO_2$$

The reaction presumably involves the intermediate, A;

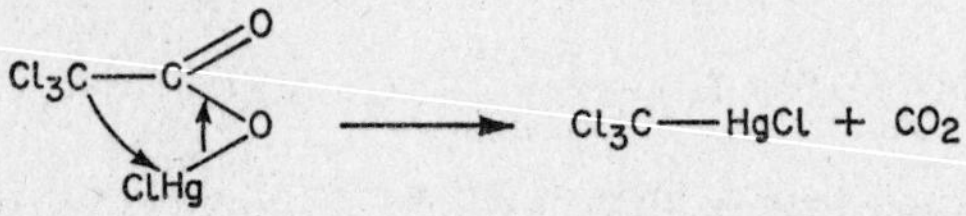

(A)

Similarly certain triphenyl-lead salts of organic acids with electron-withdrawing substituents decarboxylate on heating to about 160°C under reduced pressure, e.g.

$$Ph_3PbOCOCH_2CO_2Et \longrightarrow Ph_3PbCH_2CO_2Et + CO_2$$

A rather unusual reaction of this type is the decomposition of tributyltin formate, in which tributyltin hydride is produced:

$$Bu^n_3Sn-O-CHO \xrightarrow[10\ mm]{170°C} Bu^n_3SnH + CO_2$$
$$60\%$$

Mercuration and thallation of aromatic compounds

Aromatic compounds undergo electrophilic substitution by mercury-containing cations HgX^+. Mercuric acetate in ethanol is frequently used; benzene and toluene are mercurated smoothly by this reagent at the boiling point to mono-substituted derivatives:

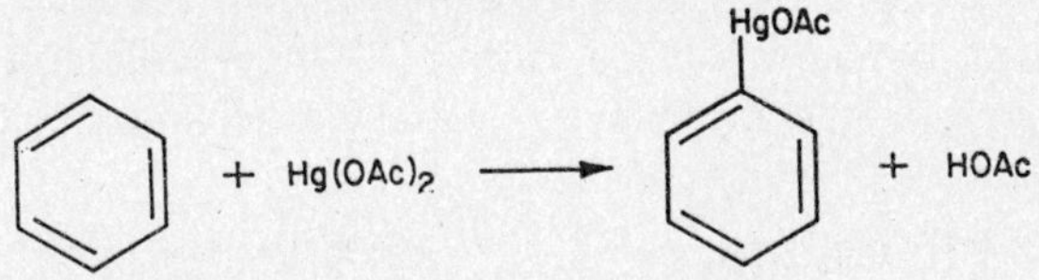

$$\text{(benzene)} + Hg(OAc)_2 \longrightarrow \text{(C}_6\text{H}_5\text{HgOAc)} + HOAc$$

Aromatic compounds which are more susceptible to electrophilic substitution than benzene are attacked more readily; thus thiophene gives a 2,5-dimercuriacetate, 2.1, and furan a 2,3,4,5,-tetramercuriacetate, 2.2, with mercuric acetate in boiling ethanol:

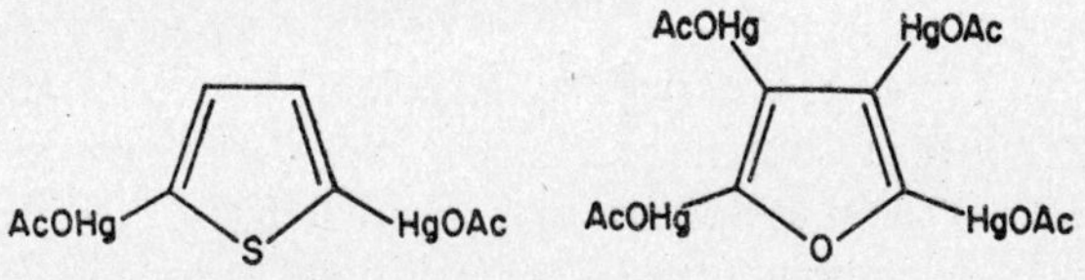

2.1 (thiophene with AcOHg and HgOAc substituents at 2,5 positions)

2.2 (furan with AcOHg and HgOAc substituents at 2,3,4,5 positions)

The preferential mercuration of thiophene provides a good method for its removal from commercial benzene.

Kinetic studies of the mercuration of aromatic compounds show that the reaction is first order both in aromatic compound and in mercuric acetate:

$$Rate \propto [ArH][Hg(OAc)_2]$$

The reaction is catalyzed by strong, weakly co-ordinating, acids (e.g. $HClO_4$), the addition of acid contributing to the catalysis up to the point where the molar ratio $HClO_4 : Hg(OAc)_2 = 1 : 1$, suggesting the formation of a fairly stable $1 : 1$ complex:

$$HClO_4 + Hg(OAc)_2 \longrightarrow [HgOAc]^+[ClO_4]^- + HOAc$$

In the presence of even a small quantity of perchloric acid, toluene is rapidly mercurated at room temperature.

Thallic salts metallate benzene and thiophene similarly, e.g.

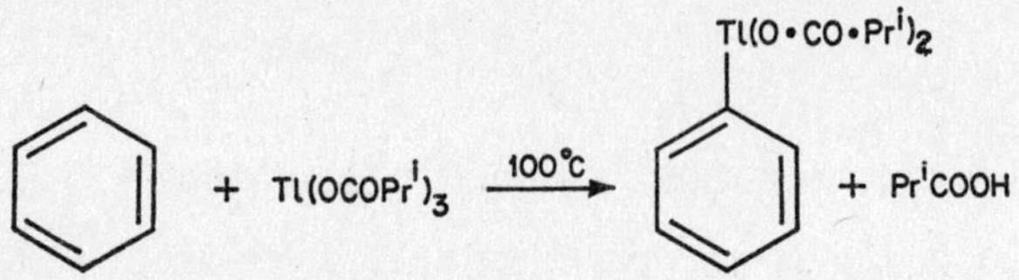

Mercuration of olefins and acetylenes

The addition of mercuric salts to olefins to yield products containing mercury-carbon bonds was discovered in 1900 by Hofmann and Sand, who obtained two series of products, 2.3 and 2.4, according to the reaction conditions.

$$HOCH_2CH_2HgCl \qquad\qquad O(CH_2CH_2HgCl)_2$$
$$\text{2.3} \qquad\qquad\qquad\qquad \text{2.4}$$

The equilibrium constant for the reaction

$$H_2O + HgCl_2 + C_2H_4 = ClHgCH_2CH_2OH + HCl$$

has been found to be $K = \dfrac{[ClHgCH_2CH_2OH][HCl]}{[HgCl_2][C_2H_4]} = 3 \cdot 0$ at 25°C. The

reaction may be driven to the right by addition of aqueous alkali during the passage of ethylene, to remove the hydrochloric acid which is formed. Proton magnetic resonance spectra of 2.3 prove that it has the structure shown containing a mercury-carbon σ–bond, and is not a π–complex as had been thought earlier. Some of its reactions, however, are not typical of mercury alkyls. For example mineral acids readily regenerate olefin rather than produce an alcohol, e.g.

$$HOCH_2CH_2HgCl + 5\%\,aqHCl \longrightarrow H_2C{=}CH_2 + HgCl_2 + H_2O$$

Contrast the reaction,

$$(C_6H_5)_2Hg + HCl = C_6H_6 + C_6H_5HgCl.$$

The regeneration of olefins with hydrochloric acid provides a method for the separation of alkenes from alkanes. The hydrocarbon mixture is shaken with aqueous mercuric acetate or chloride in the presence of base, the aqueous layer separated, and the olefins regenerated by addition of acid.

The addition of mercuric salts to olefins generally proceeds stereo-specifically *trans*. Thus cyclohexene gives the compound, 2.5, with mercuric acetate:

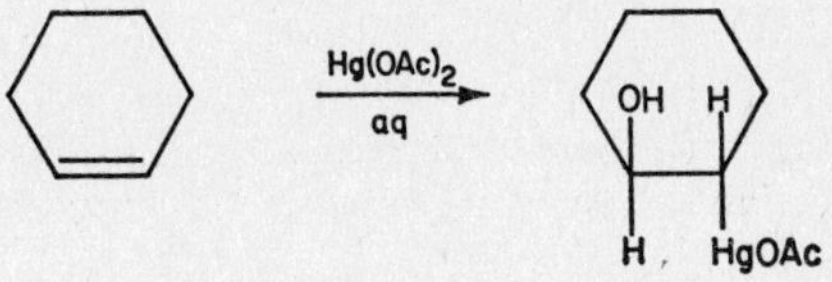

2.5

Oxymercuration/demercuration has thus been suggested as a mild procedure for the Markovnikov hydration of carbon-carbon double bonds. As shown in Chapter 3, hydroboration followed by oxidation can be used for their hydration contrary to Markovnikov's rule. The two reaction sequences are compared below for 1-hexene:

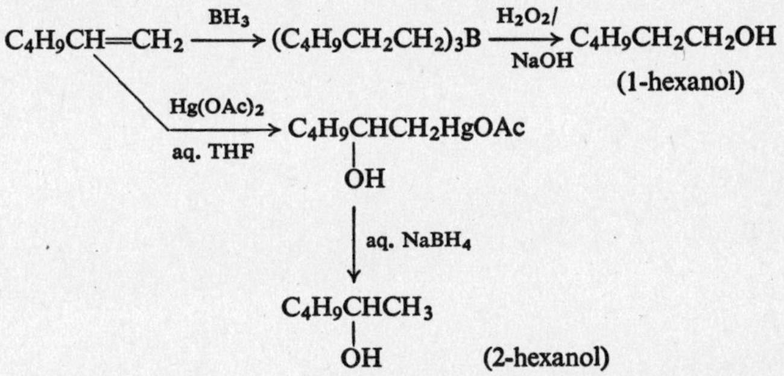

$$C_4H_9CH{=}CH_2 \xrightarrow{BH_3} (C_4H_9CH_2CH_2)_3B \xrightarrow[NaOH]{H_2O_2/} C_4H_9CH_2CH_2OH$$

(1-hexanol)

$$\xrightarrow[aq.\ THF]{Hg(OAc)_2} C_4H_9\underset{\underset{OH}{|}}{C}HCH_2HgOAc$$

$$\downarrow aq.\ NaBH_4$$

$$C_4H_9\underset{\underset{OH}{|}}{C}HCH_3$$ (2-hexanol)

Oxymercuration possesses many of the characteristics of an electrophilic addition—Markovnikov's rule is obeyed, *trans*-addition is observed, and electron-withdrawing substituents impede it.

Various compounds have been obtained by the reaction between acetylenes and solutions of mercuric salts. Thus crystalline derivatives $RC{\equiv}C{-}Hg{-}C{\equiv}CR$ are obtained from mercuric oxide and $RC{\equiv}CH$. Crystalline substances $RHgC{\equiv}CHgR$ result when RHgX reacts with acetylene itself in alcoholic potassium hydroxide. An important property of these compounds is their ready hydrolysis to aldehydes. It is probable that hydration of acetylene to acetaldehyde (industrially important some years ago) proceeds though such intermediates.

BIBLIOGRAPHY

Preparation of organometallic compounds of the main group elements

J. J. Eisch, '*Organometallic Syntheses*', Volume 2, 'Non-transition metal compounds' (Academic Press, New York, 1966). Detailed preparative procedures.

J. J. Eisch and H. Gilman, '*Advances in Inorganic Chemistry and Radiochemistry*', Volume 2 edited by H. J. Emeléus and A. G. Sharpe (Academic Press, New York, 1960). A general review.

R. G. Jones and H. Gilman, *Chem. Rev.* 54 1954, 835. Methods of preparation.

Thermochemistry

H. A. Skinner, '*Advances in Organometallic Chemistry*' edited by F. G. A. Stone and R. West (Academic Press, New York, 1964), 2, 49. Strengths of metal to carbon bonds.

CHAPTER THREE

Organometallic compounds of elements of the first three periodic groups

Introduction

In this chapter only the main group elements are considered, namely the alkali metals, Groups IIA and IIB, boron, aluminium and Group IIIB. Copper, silver, and gold are best regarded as transition elements since d orbitals participate to a substantial extent in their chemistry. Scandium, yttrium, and lanthanum are generally classified with the series of lanthanides, and the organic derivatives of these elements are few in number and do not warrant inclusion in this book. Since very little of the organic chemistry of calcium, strontium, barium, and radium is known or understood, we shall, so far as Group II is concerned, discuss little more than beryllium, magnesium and the zinc, cadmium, mercury subgroup.

Most of the classes of organometallic compounds which have found extensive use as synthetic reagents are derivatives of the elements considered in this chapter. In most cases their synthetic value is due to the polarity of the metal-carbon bond, e.g. the reactivity of the lithium alkyls and Grignard reagents is largely due to the polar $Li(\delta+)$—$C(\delta-)$ and $Mg(\delta+)$—$C(\delta-)$ bonds which very readily react with polar (or in some instances particularly polarizable) groups:

$$MeMgBr + Me_2CO \longrightarrow Me_3COMgBr$$
$$Bu^tLi + C_2H_4 \longrightarrow Bu^tCH_2CH_2Li$$

As might be expected on this basis, the relative reactivities (to most reagents) of the organic compounds of elements of the first three periodic groups run parallel to the electropositive character of the metal. Thus, the alkyls of the highly electropositive alkali metals are more reactive than those of the elements of Groups II or III. In agreement with the electronegativity increase on passing from left to right in a given Period, there is a corresponding change in reactivity of the alkyls:

$$Li > Be > B$$
$$Na > Mg > Al$$

31

General characteristics

(*i*) *The alkali metals.* Within the alkali metal group, organolithium compounds tend to differ from those of sodium and the heavier alkali metals, particularly when there is no means by which the negative charge on the carbon bound to the alkali metal can be dispersed over several other carbon atoms. Compounds such as *n*-butylsodium and *n*-butylpotassium are insoluble in saturated hydrocarbons and react with all others: their properties are generally regarded as consistent with an ionic constitution

$$(Na^+ \ \bar{C}H_2CH_2CH_2CH_3)_n$$

The intense nucleophilic reactivity of these compounds is ascribed to the presence of nearly a full unit of negative charge on one of the carbon atoms. *n*-Butyl-lithium, on the other hand, is a colourless liquid which readily dissolves in saturated hydrocarbons, as a hexamer $(Bu^nLi)_6$, believed to have an electron-deficient covalent constitution though the lithium-carbon bonds are certainly highly polar. The more covalent character of lithium alkyls is likely to be due mainly to the smaller radius and higher polarizing power of lithium relative to sodium.

It should, however, be pointed out that even if the alkali metal alkyls were all ionic, the bigger the cation and the smaller the anion the smaller would be the solubility. For example, the methyls of lithium and sodium form tetrameric aggregates $(MeM)_4$, the structure of the lithium compound being shown in Figure 9. The insolubility of methyl-lithium in hydrocarbons could well be due largely to electrostatic interaction between the positive lithium atoms of one tetramer and the negative carbon atoms of a neighbouring tetramer. Tertiary butyl-lithium also

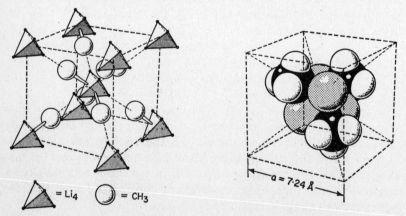

Figure 9. Unit cell of Methyl-lithium. The $(CH_3Li)_4$ tetramer. (After E. Weiss and E. A. C. Lucken, *J. Organometal. Chem.*, **2**, 1964, 197.)

forms a tetramer but, because of the greater size of the *t*-butyl relative to the methyl group, electrostatic forces *between* (as opposed to within) tetramer units would be considerably less in $(Bu^tLi)_4$ than in $(MeLi)_4$. This is why *t*-butyl-lithium is not only soluble in (for example) hexane but also is much more volatile than methyl-lithium. Conversely, keeping the size of the alkyl group constant and increasing that of the alkali metal would *increase* interaction between neighbouring aggregates. Hence, in general, sodium (and still more, potassium) alkyls should be less soluble than lithium alkyls. The more ionic character of sodium-carbon relative to lithium-carbon bonds should, of course, augment this effect.

The structure of methylpotassium is different from those of methyl-lithium and -sodium, and is consistent with an ionic constitution, K^+Me^-, each methyl group being surrounded by six potassium atoms in a trigonal prismatic arrangement.

If the negative charge on the organic radical can be dispersed over several carbon atoms (in a suitable conjugated system) then the net charge on each carbon atom becomes substantially less than unity and the nucleophilic reactivity of the carbanion is also less. This effect is apparent in a large number of organo-alkali compounds, and is usually accompanied by the appearance of vivid colours (due to $\pi \longrightarrow \pi^*$ transitions in the visible region). For example, benzylsodium (bright red) which does not attack ether is less reactive than phenylsodium (colourless) which does. It is formed when phenylsodium reacts with toluene:

$$C_6H_5^-Na^+ + C_6H_5CH_3 \longrightarrow C_6H_5CH_2^-Na^+ + C_6H_6$$

Whereas the negative charge on a phenyl anion would be mainly confined to one doubly occupied orbital of the σ–framework,

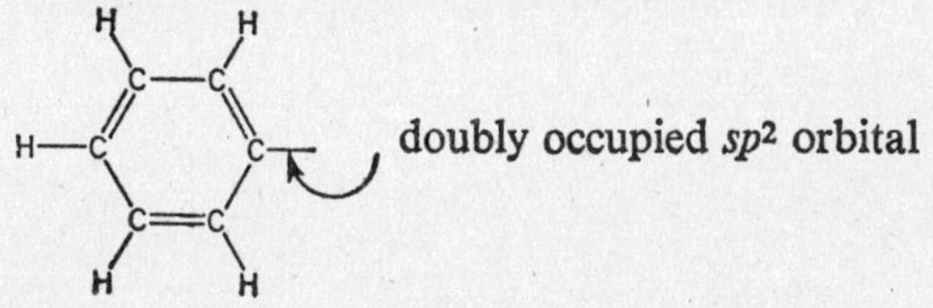

doubly occupied sp^2 orbital

the negative charge on a *benzyl* ion is distributed mainly over four carbon atoms as indicated in the following formulae:

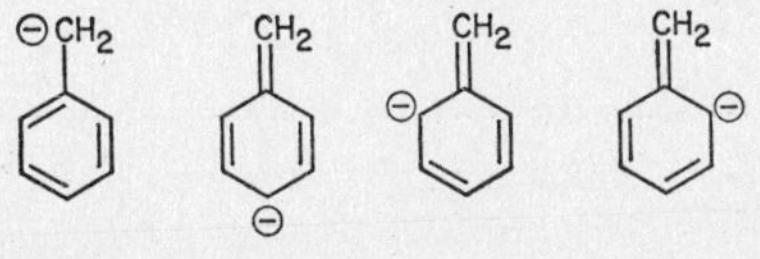

The example of alkali metal cyclopentadienides $M^+C_5H_5^-$, in which the negative charge is dispersed over all five CH groups of the aromatic anion, has already been mentioned in Chapter 1 (p 5).

A large class of charge-delocalized organoalkali compounds consists of the *addition* compounds between alkali metals and bi- or poly-nuclear aromatic hydrocarbons. Naphthalene, for example, reacts with sodium in strongly donor ethers, like tetrahydrofuran or 1, 2-dimethoxyethane, the reaction consisting of an electron transfer from the sodium to the lowest vacant π-molecular orbital of the aromatic compound. The electron transfer is normally followed by solvation of both sodium cations and hydrocarbon anions. Though these brightly coloured compounds are salts, and would exist mainly as ion-pairs (or higher aggregates) in these not very polar solvents, the alkali metal atoms should not be regarded as associated with any particular carbon atoms since the negative charge is spread over the whole of the extended π-orbital involved.

$$C_{10}H_8 + Na \longrightarrow Na^+[C_{10}H_8{}^-]$$

Addition compounds such as sodium- or lithium-naphthalene, $Li^+[C_{10}H_8{}^-]$, are to be distinguished from *substitution* compounds such as 1- or 2-naphthyl-lithium, $C_{10}H_7Li$. The formation of hydrocarbon anions is a typical property of alkali metals only on account of their relatively low ionization potentials. Aromatic hydrocarbons whose electron affinities are high also form anions on reaction with some of the more electropositive Group II metals (e.g. Mg, Ca). The subject has been very clearly reviewed (de Boer, see p 119).

(ii) *The second group metals.* The organic derivatives of these show a striking gradation of properties, general chemical reactivity in particular, running parallel to the electronegativity of the elements, as shown in Table IV.

Table IV. *Electronegativity and Reactivity, Group II Alkyls*

| Metal | Pauling Electronegativity | Ethers | Reaction of R_2M with | | | |
			CO_2	$RCOCl$	H_2O	O_2
Ba, Sr, Ca	0·9−1·0	+[a]	+	+	+	+
Mg	1·2	−	+	+	+	+
Be	1·5	−	+	+	+	+
Zn	1·6	−	−[b]	+	+	+
Cd	1·7	−	−	−[c]	+	+
Hg	1·9	−	−	−[d]	−	−

[a]Rates vary greatly. [b]RZn reacts in certain cases. [c]Reacts in presence of halide. [d]Very slow reaction.

The hydrolysis of dimethylmercury

$$Me_2Hg + H_2O \longrightarrow HgO + 2CH_4 \quad \Delta H \approx -2 \text{ kcal/mole}$$

is exothermic, so the fact that dimethylmercury and most other organo-
mercury compounds are not hydrolysed is an example of kinetic rather
than thermodynamic stability. The free energy change for the hydrolysis
would be considerably more negative than -2 on account of the formation
of two moles of gas. The hydrolysis of the alkyls of the lighter elements is
considerably more exothermic, e.g.

$$Me_2Zn + 2H_2O \longrightarrow Zn(OH)_2 + 2CH_4 \quad \Delta H = -59 \text{ kcal/mole}$$

and the dialkyls of Cd, Zn, Mg and Be are both thermodynamically and
kinetically unstable to hydrolysis. The kinetic stability of mercury alkyls
to water (and the slow hydrolysis of cadmium alkyls) is likely to be
connected with their relatively weak acceptor (Lewis acid) character, as
the reaction of the alkyls of the first three groups with water, ROH, or
RNH_2, is believed to proceed through a transition state in which the
base is co-ordinated to the metal alkyl. The decrease of reactivity with
increasing electronegativity in the sequence zinc, cadmium, mercury is
illustrated by the reaction of the dimethyls with water: dimethyl-zinc is
hydrolysed with explosive violence, -cadmium only slowly and -mercury
not at all. Effects due to the polarity of the metal-carbon bonds are
predominant in this group, relative to effects due to mean metal-carbon
bond energy. Silicon and the elements of Group IVB all have much the
same electronegativity (about $1\cdot8$), so the metal-carbon bonds all have
rather similar polarities and the dominant factor is then the decrease of
bond energy with increasing atomic number (as discussed in Chapter 1,
and resulting in the reactivity sequence $Si < Ge < Sn < Pb$).

The organic derivatives of the most electropositive elements of Group
II—Ca, Sr, Ba—have been little studied, and only a very limited amount
of reliable information is available. Organic derivatives of calcium appear
to resemble somewhat those of lithium, for example, in respect of attack
on ethers and on olefins.

Magnesium and beryllium form organic derivatives which are essentially
covalent in structure and highly reactive, the M—C bonds being strongly
polarized; they resemble organolithium compounds in several respects.

The special place occupied by the Grignard reagents is due both to
their reactivity (coupled with stability in ether solution, unlike RLi) and
the comparative ease of their preparation from metallic magnesium, which
itself is exceptional among the more electropositive metals in being readily
available and requiring little or no cleaning from corrosion products before
use.

(iii) *The third group elements*. Under this heading are considered compounds of boron, aluminium, gallium, indium and thallium, derivatives
of boron and aluminium being far more important than those of the IIIB

metals. As with the second group elements the general chemical reactivity and the acceptor character of the trialkyls relate to the electronegativity of the elements:

Reactivity and acceptor strength of Me_3M $B < Al > Ga > In > Tl$
Pauling electronegativity $2\cdot0$ $1\cdot5$ $1\cdot6$ $1\cdot7$ $1\cdot8$

The mean bond dissociation energies in the Group III trimethyls fall with increasing atomic number B, 87; Al, 66; Ga, 59. There is no great difference between the reactivities of Me_3Al and Me_3Ga, both having somewhat similar mean bond energies and the electronegativity difference being slight. In the case of boron and aluminium, however, both effects operate in the same direction and trimethylborane is much less reactive than trimethylaluminium. However, every Group III trimethyl inflames in air.

Of the main group metals and metalloids, only boron and possibly beryllium form compounds in which a substantial degree of $(p—p)\pi$ bonding is likely to occur. This property is shown particularly clearly in various boron-nitrogen compounds, but is also evident in B—O and B—X bonds. For example, the heterocyclic ring in the BN phenanthrene analogue 3.1

3.1   3.2

has clearly demonstrated aromatic character, whereas the AlN analogue is a polymer in which the heterocyclic ring has no aromatic character. The difference is mainly due to a disparity between the exponential part of the $B2p$ and $Al3p$ wave functions. For $(p—p)\pi$ bonding to be significant there has to be good overlap, and whereas the $B2p$ wave function would give fair to good overlap with $N2p$, the considerably larger $Al3p$ would not—moreover it has a node which would make matters worse. Similar considerations apply still more to the larger Group III atoms.

Organoboron compounds feature as reaction intermediates in numerous very useful synthetic processes, and aluminium alkyls represent the third biggest group of organometallic compounds made industrially (after the alkyl-lead anti-knocks and the silicone polymers).

Structural aspects of the metal alkyls

Some of the alkyls of the elements of the first three groups are covalently bound monomers, while others are associated 'electron-deficient molecules', and others may have ionic constitutions.

No monomeric alkali metal alkyls or aryls are known, as those crystal structures which have been determined indicate electron-deficient, e.g. $(MeLi)_4$, or ionic (K^+Me^-) constitutions. The dialkyls of the lighter second group metals are mostly electron-deficient dimers or polymers, but those of zinc, cadmium and mercury are monomers with a linear structure as expected from participation of one (metal) s and one p orbital (with or without d_{z^2} participation). In the third group the pattern is more complex. Whereas the trialkylboranes are monomeric, boron hydrides (and alkyl hydrides) and polyboron compounds form electron-deficient structures. Aluminium alkyls and alkyl hydrides are normally electron-deficient dimers or trimers; gallium trialkyls are monomeric though the trivinyl is a dimer; trimethylindium is a weakly associated tetramer in the solid state, otherwise all indium and thallium trialkyls appear to be monomers.

Trimethylborane has a lower boiling point $(-22°)$ than any C_4 hydrocarbon and the vapour is monomeric. Since the boron atom forms three bonds in one plane, using the $2s$ and two of the $2p$ orbitals, it has a share in only six electrons—the third $2p$ orbital being vacant (apart from possible second order hyperconjugative effects). In one sense, namely that of the electron shell falling short of the octet, trimethylborane could be described as an electron-deficient molecule; this is the same usage as the description of a carbonium ion R_3C^+ as electron-deficient. In the conventions of organometallic chemistry, however, it is more usual to reserve the term electron-deficient to the kinds of compounds described as electron-deficient at the beginning of this section, e.g. $(MeLi)_4$. The best-known example of an electron-deficient molecule in this sense is diborane, B_2H_6, in which each boron atom makes use of its $2s$ and all its $2p$ orbitals:

$$
\begin{array}{ccc}
H & \quad H \quad & H \\
\diagdown & B \cdots\cdots B & \diagup \\
H & \quad H \quad & H
\end{array}
\quad = \quad 2BH_3 \qquad \Delta H = 33\ \text{kcal/mole}
$$

Each of the hydrogen atoms in the BH_2B bridge in diborane can, taking one common view of the bonding, be regarded as bound to two boron atoms by means of a three-centre molecular orbital derived from Bsp^3 $+Hs+Bsp^3$ atomic orbitals. There are enough valence electrons (12) to provide eight for the normal terminal electron-pair B—H bonds, and two each for the two $B\cdots H\cdots B$ bridge bonds. Since each of the two three-centre molecular orbitals holds two electrons, each $B\cdots H$ bond in the BH_2B bridge could reasonably be regarded as a half-bond and is conventionally represented by a dotted line (as in the above formula of diborane). In this sense, therefore, an electron-deficient molecule is a

D

compound whose structure can best be described in terms of delocalized molecular orbitals, there being too few electrons to allow two for every link between pairs of atoms close enough to be regarded as covalently linked. Frequently, particularly in the case of bridged alkyls or hydrides, they contain a group of n atoms held together covalently by fewer than $2n$–2 electrons.

The enthalpy of one mole of diborane is less than that of two moles of BH_3 by 33 kcal, the difference being at least partly due to the greater volume available to the electron pair constituting the $B\cdots H\cdots B$ bridge relative to a single B—H (electron pair) bond. Substitution of alkyl groups for hydrogen results in dimers containing the BH_2B bridge, rather than methyl bridges (as occur in the trimethylaluminium dimer discussed below), up to the tetra-alkyldiborane stage, e.g. $Me_4B_2H_2$:

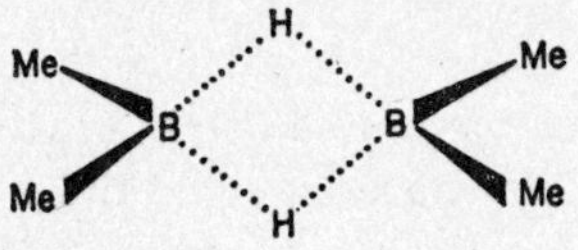

Monomeric trimethylborane has a vacant $2p$ orbital on the boron, which may interact weakly with the methyl groups. It is believed that steric reasons are the main obstacle to dimerization forming Me_6B_2. For example, the C—C distance across the four-membered ring in the dimer would be only 2·9 Å, which is very short, and the B—B distance would be still less.

Trimethylaluminium, in contrast, forms a dimer whose enthalpy is 20 kcal lower than that of two moles of monomer (both in the gas phase). The crystalline solid (mp 15°) consists of a molecular lattice of dimers, whose structure is shown in Figure 10. The aluminium atom being bigger than boron there is no steric problem, and the C—C distance across the four-membered ring is 3·41 Å. The bridging carbon atoms are five-co-ordinate.

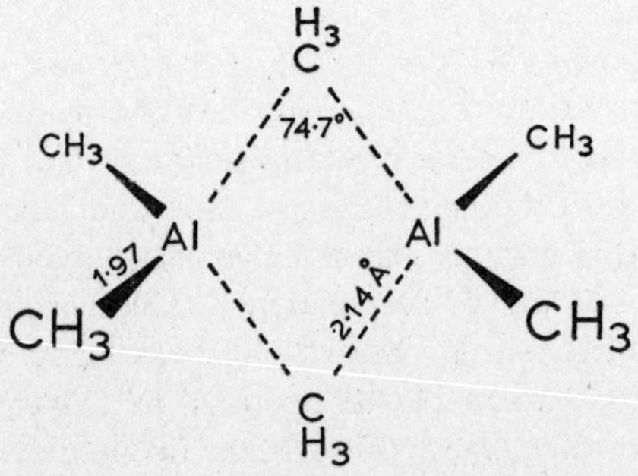

Figure 10. The trimethylaluminium dimer.

The dimerization of Me_3Al relates to the tendency for aluminium to use all four (3s + three 3p) of its available low-energy atomic orbitals. The Al—C—Al bridge bonds are commonly regarded as formed by three-centre molecular orbitals (Al sp^3 + C sp^3 + Al sp^3, see Figure 11). The rather small Al—C—Al angle, 74·7 ± 0·4°, results from the requirement of good overlap; smaller angles would result in excess Al—Al repulsion. The Al—Al distance, 2·60 Å is only a little more than twice the Al single-bond covalent radius (1·26 Å). The bridge Al—C bonds are, as expected for what amount to half-bonds, longer than the terminal Al—C bonds.

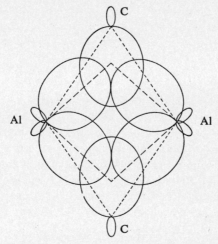

Figure 11. Schematic representation of bonding orbitals in the bridge. (After P. H. Lewis and R. E. Rundle, *J. Chem. Phys.*, 21, 1953, 986.)

Proton magnetic resonance spectra of Me_6Al_2 should distinguish between the two different kinds of methyl group. However, at room temperature only a single resonance is observed owing to the rapidity of methyl exchange, but at −75° (in solution) there are two resonances in 2 : 1 area ratio. At about −20° the mean lifetime of a particular configuration is only 3 milliseconds.

We have noted that hydrogen forms an electron-deficient bridge between boron atoms, whereas methyl groups do not. In the case of aluminium, hydrogen appears to form decidedly stronger electron-deficient bridges Al---H---Al than methyl Al---Me---Al. This is doubtless due to the non-directional character of H1s orbitals, leading to good overlap at all angles. Dialkylaluminium hydrides, 3.3, are trimeric in hydrocarbon solvents, the heat of association being 15–20 kcal/mole per hydride bridge, compared with 10 kcal./mole per methyl bridge in Me_6Al_2. The infrared absorption due to Al–H stretching is broad even for dilute solutions of $(R_2AlH)_3$, owing to the range of configurations allowed by good overlap with the Hls orbital over a range of angles. Though the dialkyl hydrides are only partly depolymerized on reaction with ethers,

$$(R_2AlH)_3 + 3R'_2O \rightleftharpoons 3R_2AlH \leftarrow OR'_2$$

(in contrast to R_6Al_2 which commonly react completely) addition of an excess of ether drives the above equilibrium to the right and the infrared Al–H absorptions of the resulting complexes are sharp—as expected for compounds with a single definite structure.

$$R_2Al \overset{\displaystyle H \cdots AlR_2}{\underset{\displaystyle H \cdots AlR_2}{\Big\langle \quad\quad \Big\rangle H}}$$

3.3

Trimethylaluminium is the most strongly associated of the aluminium trialkyls. Some dissociation is apparent at very low concentration of tri-ethyl-, tri-n-propyl-, and tri-n-butyl-aluminium in benzene, and this is ascribed to greater steric interference between bridging and terminal alkyl than in Me_6Al_2. As expected, this effect is enhanced in branched-chain alkyls, Pr^i_3Al, Bu^i_3Al and $(Me_3CCH_2)_3Al$ being monomeric. The good bridging character of methyl results in the formation of dimeric species when Me_6Al_2 reacts with Bu^i_3Al:

$$4Bu^i_3Al + Me_6Al_2 \longrightarrow 3Bu^i_2Al \overset{\displaystyle CH_3}{\underset{\displaystyle CH_3}{\Big\langle \quad\quad \Big\rangle AlBu^i_2}}$$

Since it appears that only the monomeric alkyls enter into the important reactions with olefins that are discussed below, p 93, the addition of Me_6Al_2 to other alkyls decreases their reactivity to olefins by converting them to dimers.

Alkenyl, alkynyl and phenyl groups appear more readily to act as bridging groups even than methyl. The bridging phenyl groups in Ph_6Al_2 are co-planar and the two metal atoms are symmetrically placed above and below this plane. As in the case of Me_6Al_2, the Al–C–Al angle is fairly low, 74° in one analysis and 77° in another.

Trimethylgallium is monomeric as vapour and, as shown by its infrared and Raman spectrum, also as liquid. The molecules have a planar GaC_3 skeleton of D_{3h} symmetry. It is not clear why trimethylgallium is monomeric when trimethylaluminium is dimeric: Ga and Al have similar electronegativities (1·8 and 1·6) and M—C (terminal) bond lengths

(2·0 Å). Perhaps only a small change in the thermodynamics would result in the association of Me_3Ga, and it is of interest to note that trivinylgallium is dimeric (alkenyl and alkynyl carbon bridge Al atoms better than alkyl). Triphenylgallium, however, is monomeric in hydrocarbons.

Trimethylindium is monomeric as vapour and in solution, but is a weakly associated tetramer in the crystalline state (Figure 12). Planar

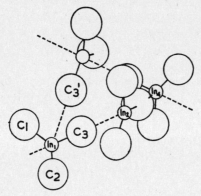

Figure 12. The trimethylindium tetramer—weak bridge bonds are indicated by broken lines. (Reproduced in modified form by permission.)

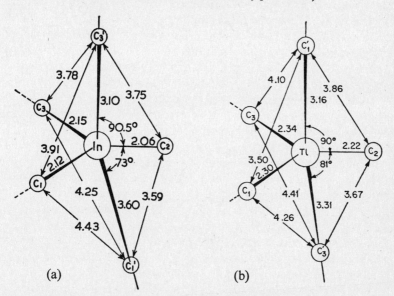

Figure 13(a). Trimethylindium arrangement of carbon atoms about indium. (After E. L. Amma and R. E. Rundle, *J. Amer. Chem. Soc.*, 80, 1958, 4141.)
      (b). Trimethylthallium. (After G. M. Sheldrick and W. S. Sheldrick *J. Chem. Soc.* (A). 1970, 28).

Me_3In units are present in the tetramer, and interaction between these is indicated by the short (3·10 and 3·60 Å) bonds between indium atoms of one Me_3In unit and carbon atoms of a neighbouring Me_3In unit (Figure 13). The Me---In—Me bridges are unsymmetrical, unlike the methyl bridge in Me_6Al_2. This exceptional type of bonding is only intelligible on the supposition that the methyl groups which take part in these bridges are somewhat flattened (in the xy plane) so that both lobes of the C $2p_z$ orbital overlap—unequally—with the vacant p orbital of neighbouring indium atoms.

Trimethylthallium (Figure 13(b)) has a similar crystal structure, though in solution both thallium trialkyls and triaryls are monomeric. A proton magnetic resonance study of Me_3Tl in $PhCD_3$ shows that methyl exchange is second-order and consistent with an electron-deficient transition state only 6 kcal/mole above that of two monomers:

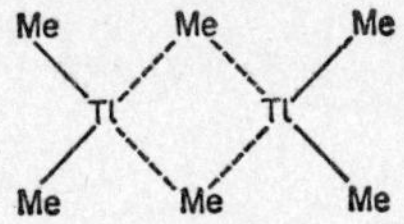

The acetylenic derivative, $Me_2TlC\!:\!CPh$, is also monomeric, though the Al, Ga and In analogues are all dimeric, with bridging alkynyl groups.

The group IIB metal dialkyls (and aryls) are all monomeric, and the metals are thus only 2-co-ordinate. Association is impeded by the too-close approach of the metal atoms that would result in the alkyl-bridged dimers, and with increase in atomic number the participation of otherwise unused p orbitals becomes less easy since the ns–np energy gap increases. The proton magnetic resonance spectra of Me_2Zn and Me_2Cd show that alkyl exchange takes place rapidly, both with each other and with alkyls such as Me_3Al known to form electron-deficient structures. Dimethylmercury exchanges only slowly.

Calcium, strontium, and barium cyclopentadienides, e.g. $(C_5H_5)_2Ca$, are known and may have ionic constitutions. The bright red bistriphenyl-methyl of calcium, $Ca(THF)_{\sim 6}(Ph_3C)_2$, appears to be a salt.

The dialkyls and diaryls of beryllium and magnesium almost without exception associate to electron-deficient structures. In contrast, both the cyclopentadienyl compounds $(C_5H_5)_2Be$ and $(C_5H_5)_2Mg$ have sandwich structures, in which the metal atoms lie between parallel C_5H_5 rings. Though the geometry is consistent with an ionic constitution, $(C_5H_5^-)_2M^{2+}$, the actual charge distributions are not known. Since the π–electron system in $C_5H_5^-$ is easily polarized, the net charges on the metal atoms are most unlikely to be as high as +2 units. The beryllium atom in $(C_5H_5)_2Be$ has been shown (by electron diffraction—crystal structure analysis gives only an average position) to be about 1·5 Å from one ring and about 2·0 Å from the other; i.e. it has two potential minima on the fivefold rotation axis. Since the beryllium atom or ion is small (radius 1 → 0·3 Å, depending on net charge), the ring-ring distance (3·37 Å) is determined mainly by repulsion between electrons in ring π-orbitals. The unsymmetrical position of the beryllium atom is responsible for the non-zero dipole moment, 2·2 D in cyclohexane.

The only other compounds R_2M (M = Be or Mg) whose structure has been firmly established at the time of writing are Me_2Be, Me_2Mg, Et_2Mg and Bu^t_2Be. The first three of these are electron-deficient polymers, and the metal atoms in the dimethyls are surrounded by four methyl groups in a nearly tetrahedral arrangement.

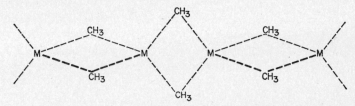

In the above formula dotted lines represent half-bonds, as in the formulae of B_2H_6 and Me_6Al_2 discussed earlier. The M—C—M angles are small (as in Me_6Al_2) to allow good orbital overlap, being 66° in $(Me_2Be)_n$ and 75° in $(Me_2Mg)_n$. The Be—C distance, 1·93 Å, is greater than the Be—C single bond length in monomeric Bu^t_2Be, 1·699 Å. Likewise the Mg—C bonds in $(Me_2Mg)_n$ are somewhat longer, 2·24 Å, than in $EtMgBr(OEt_2)_2$, 2·16 Å, which contains single Mg—C bonds. Dimethyl-beryllium and -magnesium differ in volatility. Polymers never evaporate as such, and the magnesium compound is stable to about 200° without vaporization, and is pyrolytically decomposed to hydrocarbons and carbide-like materials at higher temperatures. Dimethylberyllium, however, has a vapour pressure of 0·6 mm at 100° and 760 mm at 220°. The vapour consists of mixtures of monomeric Me_2Be (mainly) and oligomers $(Me_2Be)_{2,3...}$, the composition of the vapour depending on temperature and pressure. At low pressure gaseous Me_2Be has been shown by electron diffraction to consist of linear monomer, the Be—C distance being 1·70 Å.

Diethylmagnesium has a crystal structure basically very similar to that of Me_2Mg, i.e. it is an electron-deficient polymer containing methylene bridges. The distance between the methylene carbon of one bridge and one of the methyl carbons belonging to a bridging ethyl group in the next four-membered ring is only 3·76 Å, and this is less than the normal distance of closest approach between non-bonded carbon atoms (ca. 4 Å) by an amount which implies some appreciable strain. If, in the Et_2Mg polymer, the magnesium atoms were replaced by the smaller beryllium atoms, then this CH_2---CH_3 distance would be reduced to 3·19 Å (even for a BeCBe angle of 60°), which is much too short. This is why beryllium dialkyls containing two or more carbon atoms in each alkyl group are dimers, not polymers, examples being $(Et_2Be)_2$, $(Pr^i_2Be)_2$ and $(Bu^n_2Be)_2$. In a dimer each beryllium atom would be bound to only three alkyl groups (only two of the three $2p$ metal orbitals being used), and steric interaction

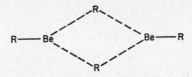

between these would be less than in polymers in which each Be would be bound to four alkyl groups. In di-t-butylberyllium, steric effects prevent the formation even of dimers, and this is the only known beryllium dialkyl which contains 2-co-ordinate (sp) beryllium at room temperature. The

$$Me_3C \xrightarrow[1·70]{} Be \xrightarrow[1·70]{\overset{\overset{180°}{\frown}}{}} CMe_3$$

bulky $(Me_3Si)_2N$–group is also big enough to result in $[(Me_3Si)_2N]_2Be$ being monomeric.

There is less structural information about the alkali metal alkyls than about organic compounds of the second and third groups, and at the same time there is much scope for the formation of relatively complex structures. The Group II and III metal alkyls, when monomeric, have two and one vacant p orbitals respectively, and, as already discussed, associate to form electron-deficient dimers or polymers in which the bridging carbon atoms are 5- and the metal atoms are 4-co-ordinate. Monomeric alkali metal alkyls are unknown, but would have three vacant p orbitals. The evidence available indicates a tendency for the alkali metal alkyls to associate in such a way as to allow the participation in the metal-carbon bonding of two or more usually three p orbitals. Crystalline methyl-lithium consists of tetramers, whose structure has already been mentioned (Figure 9, p 32). It can be seen that each carbon atom is bound to three hydrogen atoms (the CH_3 group) and has three lithium atoms as equally near neighbours; thus the co-ordination number of the carbon is at least six. The structure consists of two interpenetrating and unequal tetrahedra, one Li_4, one C_4, but the Li_4C_4 skeleton can be represented as a distorted cube. The shortest distances (other than the CH bonds) are between lithium and carbon, 2·28 Å, and the principal bonding holding the tetramer together is thus likely to be between lithium and carbon. Each lithium and each carbon can be considered as contributing one electron, hence eight electrons hold the Li_4C_4 skeleton together. One molecular orbital (the lowest in energy) is derived from a symmetrical combination of the four Li $2s$ and the four C $2s$ atomic orbitals; this will accommodate two electrons. Three other bonding molecular orbitals, equal in energy, can be derived from the Li $2p_x$ and C $2p_x$, Li $2p_y$ and C $2p_y$, and Li $2p_z$ and C $2p_z$ atomic orbitals. These three molecular orbitals accommodate the remaining six electrons. Other, antibonding, molecular orbitals can be constructed, but these hold no electrons when the molecule is in its ground state. On the above basis the carbon is six- and the lithium three-coordinate. However the Li-Li distances are 2·56 Å, which is shorter than in lithium metal (3·04) or in Li_2 (2·67 Å), and there may be some Li—Li bonding arising from a still more delocalized system of bonds. The methyl carbon of one tetramer is close (2·54 Å, see Figure 9) to a lithium atom of a neighbouring tetramer. It is this interaction which makes methyl-lithium involatile and sparingly soluble in hydrocarbons.

Ethyl-lithium has a more complicated structure, the lower symmetry of the ethyl groups resulting in linear (instead of three-dimensional) association of tetramer units. t-Butyl-lithium is tetrameric in solution, and a tetrahedral structure like that of methyl-lithium is indicated by its infrared and Raman spectra. The bulk of the t-butyl group greatly reduces interaction between tetramers, and $(Bu^tLi)_4$ sublimes at 70°/0·1 mm and is much more volatile than $(MeLi)_4$.

Ethyl-lithium, though consisting basically of tetramer units associated into strips by tetramer-tetramer interactions in the crystal, both dissolves in hydrocarbons in hexameric form $(EtLi)_6$, and also vaporizes (from a study of its mass spectrum) as a mixture of tetramer and hexamer. The much-used reagent, n-butyl-lithium, is also hexameric in hydrocarbon solution, but the structure of the hexamer is not definitely known. Menthyl-lithium is dimeric both in benzene and in cyclohexane, and this may well be connected with its considerably higher reactivity (e.g. to bromo-benzene) than, say, $(Bu^nLi)_6$. The state of lithium alkyls in donor (e.g. ether) solvents is mentioned at the end of this chapter.

Preparative aspects

The alkali metals: [A] lithium

The lithium alkyls are extremely reactive, being sensitive to oxygen and to moisture. Care is therefore needed to use dry apparatus, *dry solvents* and to exclude air by means of dry oxygen-free nitrogen. Generally, magnetic stirrers are preferred to other types which may allow air in through the stirrer-gland.

(*i*) *Lithium metal and halides.* When organolithium compounds are needed for use as synthetic reagents they are commonly obtained by the reaction:

$$2Li + RX \longrightarrow RLi + LiX$$

One of the most commonly used reagents is *n*-butyl-lithium, and if a laboratory has more than occasional use for this, it is probably best to buy it ready-made as a solution (about 2M, reckoned as monomer) in hexane or heptane. The reaction between lithium and halides can be carried out in light petroleum, hexane, benzene or ether.

Methyl chloride, -bromide, and -iodide all react satisfactorily (ether solvent), an advantage of the first being the low halide content of the product, LiCl being only sparingly soluble. *Aryl* chlorides are often insufficiently reactive.

The physical state of the lithium is important, as reaction does not readily start if the metal is coated with much corrosion product. It is now possible to buy dispersions of very finely divided lithium in thick hydrocarbon oil or wax. The oil can be removed by washing with ether or hexane, on which the lithium floats. The presence of small amounts of sodium sometimes has a big effect on the yields of RLi from metal and RX, and it is fortunate that commercial lithium normally contains a suitable amount of sodium as impurity.

(*ii*) *Transmetalation.* Metal-metal exchange reactions are often quite fast at room temperature, particularly when at least one of the metals readily forms electron-deficient compounds (as lithium certainly does). Some of these transmetalations provide convenient preparative processes, particularly when one component of an equilibrium is sparingly soluble. For example, vinyl-lithium, when required in solution for a subsequent vinylation reaction, can conveniently be obtained from PhLi (itself prepared from PhBr + 2Li) and commercially available tetravinyltin in diethyl ether:

$$4PhLi + (CH_2:CH)_4Sn \longrightarrow 4CH_2:CHLi + Ph_4Sn \downarrow$$

Only about a gram of Ph_4Sn dissolves in a litre of ether at room temperature. Allyl-lithium can be made similarly, in solution using ether as solvent,

$$(CH_2:CH \cdot CH_2)_4Sn + 4PhLi \longrightarrow 4CH_2:CH \cdot CH_2Li + Ph_4Sn \downarrow$$

or as a pyrophoric precipitate using hydrocarbon,

$$(CH_2:CH \cdot CH_2)_4Sn + Bu^nLi \longrightarrow (CH_2:CH \cdot CH_2)_3SnBu^n + CH_2:CH \cdot CH_2Li \downarrow$$

Reactions between metallic lithium and organic compounds of more electronegative metals are not properly classed as transmetalations, but are mentioned at this point. An example is the reaction between lithium and tetravinyl-lead, catalyzed by a trace of benzophenone:

$$(CH_2:CH)_4Pb + 4Li \xrightarrow{Et_2O} 4CH_2:CHLi + Pb \downarrow$$

The reaction between diphenylmercury and lithium has often been used to obtain halide-free phenyl-lithium,

$$Ph_2Hg + Li(excess) \longrightarrow 2PhLi + Li, Hg$$

but phenyl-lithium is more conveniently obtained by a metal-halogen exchange as discussed below.

(iii) *Metal-halogen exchange.* This is the reaction

$$RLi + R'X \rightleftharpoons R'Li + RX$$

which is generally fast at room temperature or below. As indicated in Table V below, the metal becomes preferentially attached to the more electronegative organic radical, or the radical which would form the more stable carbanion.

Table V. *Lithium-halogen exchange equilibria*
$$RLi + PhI \rightleftharpoons RI + PhLi^a$$

R	log K	R	log K
vinyl	$-2 \cdot 4$	isobutyl	$4 \cdot 6$
cyclopropyl	$1 \cdot 0$	neopentyl	$5 \cdot 5$
ethyl	$3 \cdot 5$	cyclobutyl	$6 \cdot 1$
n-propyl	$3 \cdot 9$	cyclopentyl	$6 \cdot 9$

^{a}In ether, or 40% ether 60% pentane, at $-70°$

Since *n*-butyl-lithium is readily available it is a very suitable source of *aryl*-lithium compounds, since aryl groups are substantially more electro-

$$Bu^nLi + ArX \longrightarrow Bu^nX + ArLi$$

negative than *n*-butyl. This has been a very widely used route to aryl-lithium compounds, and numerous aryl halides which do not react satisfactorily with lithium afford lithium aryls in high yield on reaction with butyl-lithium, e.g.

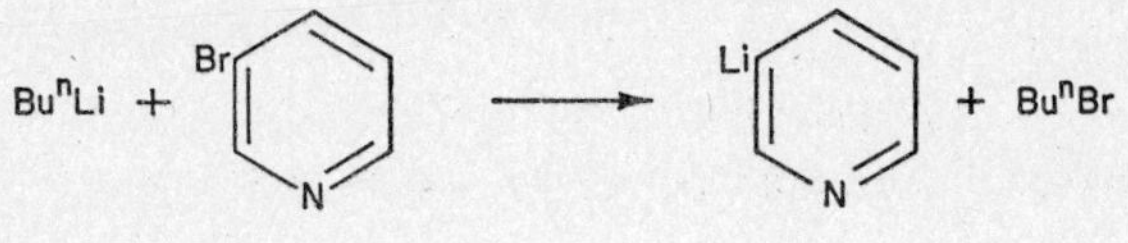

The preparation of crystalline phenyl-lithium is an interesting and recent application

$$Bu^nLi + PhI \xrightarrow{C_6H_6} Bu^nI + PhLi \downarrow$$

(*iv*) *Metal-hydrogen exchange* (*metalation*). This is the reaction between an organolithium compound, RLi, and a hydrocarbon R′H which is more acidic than RH. A simple example is:

$$PhLi + PhC\vdots CH \longrightarrow C_6H_6 + PhC\vdots CLi$$

phenylacetylene being much more acidic than benzene. As alkanes are less acidic than aromatic hydrocarbons and aromatic heterocyclics, lithium derivatives of the latter can often be obtained by reaction with butyl-lithium.

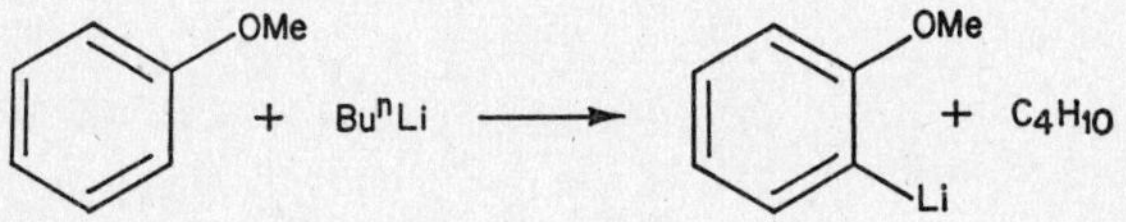

The course of such reactions is less predictable than with metal-halogen exchange, since the question of *orientation* has to be considered. Metalation is believed to involve attack by the nucleophilic (hydrocarbon) part of the organolithium reagent on a hydrogen atom of the molecule being metalated. That anisole and $PhCF_3$, for example, are both metalated mainly in the *ortho* position is due to the $-I$ inductive effect of the $-OMe$ and $-CF_3$ groups making the *ortho* hydrogen more acidic. Similarly, an aliphatic hydrocarbon radical deactivates the *ortho* hydrogen atoms, and isopropylbenzene, though not metalated by butyl-lithium in ether, is metalated by sodium and potassium alkyls in the *meta* and *para* positions.

The nucleophilic character of carbon bound to lithium appears to be increased when the lithium is co-ordinated to a base. Thus metalation of a given hydrocarbon by butyl-lithium increases with the base strength of the solvent: hydrocarbon < ether < tetrahydrofuran < tertiary amines. The difficulty about using ethers (particularly THF) as solvent is that of avoiding metalation of the ether α to the oxygen:

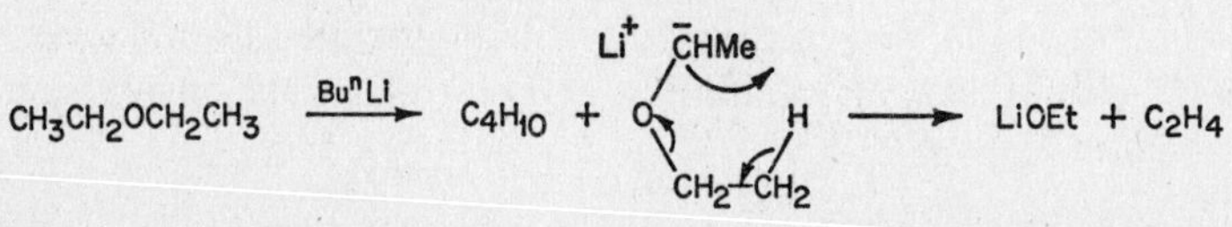

This difficulty does not arise significantly with tertiary amines, since CH bonds adjacent to nitrogen in tertiary amines are less susceptible to nucleophilic attack than those adjacent to oxygen in ethers.

The n-butyl-lithium chelate complex with tetramethylethylenediamine (TMED) is a particularly powerful metalating agent

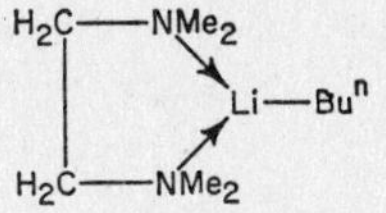

$$\begin{array}{c} H_2C\text{----}NMe_2 \\ | \qquad\qquad Li\text{---}Bu^n \\ H_2C\text{----}NMe_2 \end{array}$$

This monomeric complex is very soluble in paraffin solvents and the lithium-carbon bond is so strongly polarized that it may be regarded as the best source of highly reactive soluble carbanions currently available. It metalates toluene at room temperature, giving the TMED complex of benzyl-lithium

$$Bu^nLi(TMED) + PhCH_3 \longrightarrow C_4H_{10} + PhCH_2Li(TMED)$$

Ferrocene is also metalated, and after reaction with CO_2 gives the dicarboxylic acid in high yield.

Several types of metalation reaction are important because they result in the formation of synthetically useful reaction intermediates. One example is the metalation of CH bonds rendered acidic by proximity to a P^+ atom in a phosphonium ion, giving *Wittig reagents*, e.g.:

$$Ph_3P^+CH_3 + RLi \ [\longrightarrow Ph_3P^+CH_2Li + RH] \longrightarrow Ph_3P{=}CH_2 + Li^+$$

An intermediate containing a C—Li bond may not be involved, as the reagent also may be formed directly by proton abstraction from the phosphonium ion. Wittig reagents react under mild conditions with carbonyl compounds

$$Ph_3P{=}CH_2 + R'_2CO \longrightarrow Ph_3P{:}O + R'_2C{:}CH_2$$

In some instances an intermediate product has been isolated:

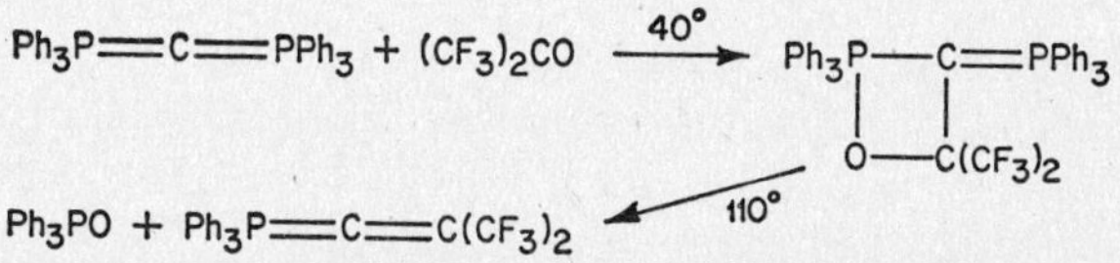

$$Ph_3P{=}C{=}PPh_3 + (CF_3)_2CO \xrightarrow{40°} \begin{array}{c} Ph_3P\text{---}C{=}PPh_3 \\ | \qquad\quad | \\ O\text{---}C(CF_3)_2 \end{array}$$

$$Ph_3PO + Ph_3P{=}C{=}C(CF_3)_2 \xleftarrow{110°}$$

The metalation of *gem*-dichlorides may result in the transient formation of *carbenes*, which may then react further forming cyclopropanes and other products.

$$CH_2Cl_2 + Bu^nLi \longrightarrow LiCHCl_2 + C_4H_{10}$$

$$LiCHCl_2 \longrightarrow LiCl + \ddot{C}HCl$$

The reactions of Wittig reagents, carbenes and benzynes (which can also be formed by metalations) are outside the scope of this book: the reader is referred to texts on organic chemistry.

(*v*) *Analysis.* Much attention has been given to methods for analyzing organolithium reagents. Not only are the more usual preparative methods (e.g. $RX + Li$) not quantitative, but the reactivity of the product to air and often to the solvent used makes analysis essential before the reagent is used. In cases where attack on ether is slow and accidental oxidation (giving ROLi) has definitely been excluded, then analysis by hydrolysis and subsequent titration of total alkali may be sufficient:

$$RLi + H_2O \longrightarrow RH + LiOH$$

Freshly prepared phenyl- and methyl-lithium in ether can normally be analyzed satisfactorily in this simple way. Attack of RLi on diethyl ether at a given concentration and temperature increases in the order $R = Me < Ph < Bu^n < Et < Bu^s < Bu^t$. If, for example, *n*-butyl-lithium has been prepared in diethyl ether, and maybe kept overnight, then some appreciable attack on the solvent is likely to have taken place:

$$Bu^nLi + Et_2O \longrightarrow C_4H_{10} + C_2H_4 + EtOLi$$

Hydrolysis would then give alkali arising both from Bu^nLi and EtOLi, so the titration with acid would overestimate the amount of Bu^nLi present. Several methods have been devised for dealing with this problem. In one of these a weighed amount of benzoic acid in a mixture of dimethyl-sulphoxide, 1,2-dimethoxyethane and ether or hydrocarbon is titrated with the organolithium solution. The indicator is triphenylmethane, which develops a red colour when all the benzoic acid is used up:

$$RLi + PhCO_2H \longrightarrow PhCO_2Li + RH$$
$$RLi + Ph_3CH \longrightarrow RH + Ph_3CLi \text{ (red)}$$

The described procedure (see Bibliography) also copes satisfactorily with the problem of water adsorbed on the surface of the reaction vessel. Another method makes use of the formation of coloured complexes between 1,10-phenanthroline and reactive organometallic compounds such as RLi. A trace of phenanthroline is added to, for example, butyl-lithium producing a rust-red colour. The mixture is then titrated with *sec*-butanol in xylene (used because $LiOBu^s$ is soluble in xylene) until the colour is just discharged. References are given on p 119. Since *n*-butyl-lithium is a much-used reagent and is occasionally stored in the laboratory for appreciable periods, it should be noted that lithium *n*-butoxide, though insoluble in hydrocarbons, is freely soluble in hydrocarbon solutions of butyl-lithium. Thus the absence of a precipitate in stored butyl-lithium solutions is no proof that oxygen has been excluded.

(*vi*) *Some applications.* The use of organolithium compounds in forming Wittig reagents, benzynes and carbenes has been touched on already. Here are mentioned a few reactions with olefins.

n-Butyl-lithium adds to $Ph_2C:CH_2$ giving a coloured (charge-delocalized) product, which can be written as an ion-pair (at least for donor solvents):

$$Bu^nLi + Ph_2C:CH_2 \longrightarrow Bu^nCH_2Ph_2C^-Li^+$$

The reaction is more complex than suggested above, since Bu^nLi is hexameric in hydrocarbons and tetrameric in ether. This reaction, and most other instances of RLi adding to olefin, go faster in ethers than in hydrocarbons, and fastest of all in the presence of tertiary amines. Primary lithium alkyls do not add to ethylene in ether, but secondary and tertiary alkyls react readily. This was discovered by chance: addition of CO_2 at $-50°$ to Pr^iLi which had been prepared and kept at $-50°$ resulted in the expected products Pr^i_2CO and Pr^iCO_2H. However, when Pr^iLi in ether was allowed to warm to room temperature an exothermic reaction set in and the mixture boiled spontaneously. Carbonation then gave $(Pr^iCH_2CH_2)_2CO$ and $Pr^iCH_2CH_2CO_2H$. This happened because Pr^iLi attacked the solvent

$$Pr^iLi + Et_2O \longrightarrow C_3H_8 + LiOEt + C_2H_4$$

and the ethylene then added to unreacted Pr^iLi

$$Pr^iLi + C_2H_4 \longrightarrow Pr^iCH_2CH_2Li \xrightarrow{CO_2} (Pr^iCH_2CH_2)_2CO + Pr^iCH_2CH_2CO_2H$$

The polarity of the Li—C bond is so much increased by co-ordination with TMED that the TMED n-butyl-lithium complex adds ethylene under moderate pressure giving polymers. Reaction in the presence of benzene gives phenyl-ended telomers:

$$Bu^nLi\text{-}TMED + C_6H_6 \longrightarrow PhLi\text{-}TMED + C_4H_{10}$$

$$PhLi\text{-}TMED + nC_2H_4 \xrightarrow[150-1000\text{p.s.i.}]{60-130°} Ph(C_2H_4)_nLi\text{-}TMED$$

$$Ph(C_2H_4)_nLi\text{-}TMED + C_6H_6 \longrightarrow PhLi\text{-}TMED + Ph(C_2H_4)_nH$$

In this process n may range from 1 to 100, and over 500 g telomer may be obtained per gram of Bu^nLi.

Reactions between RLi and *conjugated* diolefins can result in synthetic rubbers:

1,4 cis
Hevea rubber

isoprene

1, 4 trans
gutta percha

The alkali metals: [B] sodium

(*i*) *Substitution compounds.* Organosodium compounds in which the negative charge is localized, or nearly so, on one carbon atom are obtained by three main routes, $R_2Hg + Na$, $RLi + Bu^tONa$, and $RX + 2Na$.

The mercury method is carried out in light petroleum

$$R_2Hg + Na \text{ (excess)} \longrightarrow 2RNa \downarrow + Na, Hg \text{ amalgam}$$

The white insoluble products are violently reactive, and slowly metalate even paraffin hydrocarbons.

The new sodium t-butoxide method appears to be better, and works because $LiOBu^t$ is soluble and RNa insoluble in paraffin solvents:

$$RLi + Bu^tONa \longrightarrow RNa \downarrow + LiOBu^t \text{ (soluble)}$$

The reaction between sodium and halides suffers from complications due to various exchange and coupling reactions, which can lead to a variety of products. One example, however has been developed into a satisfactory preparative method which may well find industrial application. This is the reaction between chlorobenzene and fine sodium *dispersions* (0·5–20 microns) in a hydrocarbon medium.

$$PhCl + 2Na \longrightarrow PhNa + NaCl$$

Phenylsodium, which can be made continuously in one variation of this reaction, has several attractions as a phenylating agent relative to phenyl-lithium or phenylmagnesium bromide, since both chlorobenzene and sodium are cheap (compared with bromobenzene and lithium) and Grignard reagents also need expensive solvents.

Methyl-sodium (and –potassium) is more stable thermally than the higher alkyls, which decompose slowly at room temperature and rapidly at ca. 100° by loss of olefin, e.g.:

$$C_2H_5Na \longrightarrow C_2H_4 + NaH$$

and also by self-metalation processes,

$$2C_2H_5Na \longrightarrow C_2H_4Na_2 + C_2H_6 \text{ etc.}$$

Hydrogen, ethane and acetylene are formed when the residue is hydrolyzed. Sodium alkyls are, of course, very powerful metalating agents.

Organosodium compounds in which the negative charge is delocalized may be obtained by a greater variety of methods. They are all brightly coloured, and their reactivity is less. They are also commonly soluble in ether, and not normally decomposed by it, the sodium ion being solvated. Only a few fairly typical examples can be given here.

Hydrocarbons containing two or three aryl groups bound to one carbon atom are sometimes sufficiently acidic to lose a proton to the sodium salt of a strong base such as NH_2^-

$$Ph_3CH + Na^+NH_2^- \longrightarrow Ph_3C^-Na^+ + NH_3$$

A rather special case arises when the removal of a proton creates an aromatic system. Cyclopentadiene, which is not aromatic, gives sodium cyclopentadienide (which is) by reaction with sodium in tetrahydrofuran or 1, 2-dimethoxyethane, or, at lower cost for big scale preparations, with a sodium dispersion in ether-hydrocarbon mixtures.

$$3C_5H_6 + 2Na \longrightarrow 2C_5H_5{}^-Na^+ + C_5H_8$$

Triphenylmethylsodium, one of the most extensively studied members of this class of compound, is conveniently obtained, as a deep red ether solution, from the chloride and sodium amalgam.

$$Ph_3CCl + Na(Hg) \longrightarrow Ph_3C^-Na^+ + NaCl.$$

It is, of course, hydrolysed by water

$$Ph_3C^-Na^+ + H_2O \longrightarrow Ph_3CH + NaOH$$

and is oxidized by O_2 giving various peroxy compounds. Aldehydes and ketones react like hydroxy compounds regenerating Ph_3CH if enolization can occur (e.g. acetone), otherwise addition to the carbonyl group takes place:

$$PhCHO + Ph_3C^-Na^+ \longrightarrow PhCH(ONa)CPh_3$$

In certain cases an ion-radical transfer equilibrium may be realized, sometimes with striking colour changes:

$$Ph_3C^-Na^+ + Ph_2CO \longrightarrow Ph_3C^{\cdot} + Ph_2CO^{-\cdot}Na^+$$
$$\text{red} \quad \text{colour-} \qquad \text{yellow} \qquad \text{blue}$$
$$\text{less}$$

In this instance the equilibrium lies to the right in ether at room temperature and a green colour results.

(ii) Addition compounds. All the organosodium compounds discussed so far can be considered as derived from hydrocarbons by the substitution of sodium for hydrogen, e.g. $Ph_3CH \longrightarrow Ph_3CNa$. Various organosodium compounds may also be formed by *addition* of sodium to hydrocarbons: these are of two types, those that could also be regarded as sodium salts of hydrocarbons and those that cannot. For example, 1,1-diphenylethylene *adds* sodium by a one-electron transfer, and this is followed by dimerization:

$$Ph_2C{:}CH_2 + Na \xrightarrow{\text{Et}_2O} Ph_2\bar{C}{\cdot}\dot{C}H_2\}Na^+$$

$$2Ph_2\bar{C}{\cdot}\dot{C}H_2\}Na^+ \longrightarrow Ph_2\bar{C}{\cdot}CH_2CH_2{\cdot}\bar{C}Ph_2\}Na_2{}^{++}$$

The product is the disodium salt of 1,4-tetraphenylbutane (which is formed on adding water) and could thus be regarded as a substitution compound, but is commonly classified as a product of the *dimerizing addition* of sodium to an olefin. Another example of dimerizing addition (though

E

lithium has been used in this instance) is the formation of a tetraphenyl-butadiene dianion from PhC:CPh

$$PhC\!:\!CPh + Li \longrightarrow Li^+Ph\bar{C}\!:\!\dot{C}Ph \longrightarrow LiCPh\!:\!CPh\!\cdot\!CPh\!:\!CPhLi$$

Hydrolysis yields 1,2,3,4-tetraphenylbutadiene, but reaction with some dihalides has allowed some interesting heterocyclic syntheses, e.g. a highly reactive potentially antiaromatic borole.

$$LiCPh\!:\!CPh\!\cdot\!CPh\!:\!CPhLi + PhBBr_2 \longrightarrow$$

The other kind of addition compound, exemplified by sodium-naphthalene, cannot be regarded as a substitution product, as it is formed by the addition of one or more electrons to the lowest vacant molecular orbital of an aromatic hydrocarbon. Aromatic hydrocarbons containing two or more aromatic rings, joined (biphenyl, terphenyls), conjugated (1,4-diphenylbutadiene), or fused (naphthalene, anthracene), react with alkali metal *without* loss of hydrogen. These addition compounds are all strongly coloured, and their formation is greatly facilitated in basic solvents such as tetrahydrofuran or 1,2-dimethoxyethane.

Solutions of sodium-naphthalene in tetrahydrofuran are dark green, electrically conducting because the compound is a salt $Na^+(THF)_nC_{10}H_8{}^-$, and paramagnetic because of the extra electron which is in a singly occupied π-orbital. Information about the distribution of the unpaired electron about its various possible positions in the anion can in suitable cases be derived from the electron spin resonance spectrum. If the orbital occupied by the unpaired electron in a hydrocarbon anion is non-degenerate, as is the case with naphthalene and anthracene, then a *second* electron (formation of anion^{2-}) would enter the same orbital and both the paramagnetism and e.s.r. spectra disappear.

Aromatic hydrocarbons vary considerably in their electron affinities, e.g. benzene < biphenyl < naphthalene < phenanthrene < pyrene < anthracene. Addition of sodium-biphenyl, for example, to a solution of a hydrocarbon of greater electron affinity results in electron transfer, frequently accompanied by a colour change:

$$biphenyl^- + pyrene \longrightarrow biphenyl + pyrene^-$$
$$\text{green} \qquad\qquad\qquad\qquad \text{brown}$$

Such electron exchange reactions have been studied by potentiometric titration of a series of hydrocarbons using sodium-biphenyl (in THF or $MeOC_2H_4OMe$) and platinum electrodes. A typical titration is repre-

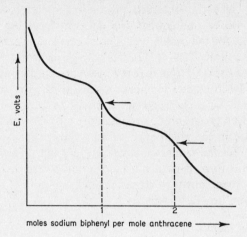

Figure 14. Potentiometric titration of anthracene with sodium biphenyl.

sented in Figure 14 which illustrates the successive formation of anthracene$^-$ and anthracene^{2-}

Solutions of sodium-naphthalene and similar compounds are very strong reducing agents, behaving to some extent as solutions of electrons. Halogen present in an organic halide is reduced to halide ion, e.g.

$$C_{12}H_{10}{}^- + RX \longrightarrow C_{12}H_{10} + R^{\cdot} + X^-$$

The radicals produced may react with molecules of solvent, hydrocarbon, hydrocarbon anion or may dimerize. Reaction with methyl iodide and subsequent determination of iodide anion in the aqueous extract is a recommended method for analyzing compounds like sodium-naphthalene. The etching of polytetrafluorethylene is an interesting example of the reduction of halides: an active surface is produced which is then able to form strong bonds to an epoxy resin. The reducing action of hydrocarbon anions has been used in the preparation of metal carbonyls. In this it is commonly convenient to start from the salt of a metal in a $+2$ or $+3$ oxidation state, which must be reduced to the zero oxidation state: both sodium-naphthalene and aluminium alkyls have been used in this connection.

The non-aromatic non-planar hydrocarbon cyclo-octatetraene adds two electrons on reaction with potassium in tetrahydrofuran, forming a *planar* aromatic dianion, The equilibrium

$$2C_8H_8{}^- \rightleftharpoons C_8H_8 + C_8H_8{}^{2-}$$

lies far to the right, and the concentration of paramagnetic ions, though big enough to be detected, is still very small ($< 10^{-3}$M in a 0·6 M solution of $K_2C_8H_8$) since the observed proton magnetic resonance spectrum would otherwise be obliterated. The dianion obeys the Hückel $(4n+2)$ π-electron rule, for the case $n = 2$. The planar and aromatic character of the colourless salt $K_2C_8H_8(THF)$ has also been deduced from its ultra-violet, infrared, and proton magnetic resonance spectra.

2,2'-Bipyridyl in ether adds alkali metals giving, for example, the deep red lithium salt, Li^+bipy^-, and in tetrahydrofuran further addition takes place giving deep green solutions containing $bipy^{2-}$ anions. These reagents have been used in the preparation of an extensive series of complexes of metals in low formal oxidation states, e.g. Cr bipy$_3$ (diamagnetic). Such compounds are formed even by main group metals, and, for example, Be bipy$_2$ has been shown by its e.s.r. spectrum to be a derivative of the bipyridyl anion, i.e., $Be^{2+}(bipy^-)_2$.

The Group IIA metals

(i) *Calcium, strontium, and barium.* Very little is known about the organic compounds of these metals. The only properly characterized derivatives are the evidently ionic cyclopentadienyls $(C_5H_5)_2M$, the complex ethyls $M(AlEt_4)_2$ (M = Ca, Sr, Ba) and the red complex $Ph_3CCaCl(THF)_2$ (from calcium amalgam and Ph_3CCl in THF). Some alkylcalcium halides have also been obtained in solution in ether and in THF.

(ii) Magnesium

(a) *Grignard reagents.* The organomagnesium halides, $RMgXL_{ca.2}$ (L = base such as diethyl ether), known as Grignard reagents are much more extensively used than the dialkyls or diaryls R_2Mg. They are named after Victor Grignard (1871–1935), who developed their chemistry in 1900 and the following years. A few magnesium dialkyls had been prepared with some difficulty as early as 1866, but Grignard's contribution, for which he was awarded the Nobel prize for Chemistry in 1912, was the discovery that organomagnesium halides could be prepared, and used for a wide variety of syntheses, from an organic halide and metallic magnesium in a basic solvent such as diethyl ether.

Grignard reagents are normally prepared by slow addition of the halide (the choice of halide has been discussed in Chapter 2) to a stirred suspension of magnesium turnings, using dry apparatus and reagents. Reaction cannot start until the reagents have penetrated in some places the thin oxide film (or imperfections in it) which coats all magnesium which has been exposed to air. The formation of Grignard reagent is strongly exo-thermic, and though the reaction is often slow to begin it commonly accelerates very markedly when an appreciable amount has been formed. Care therefore is necessary to avoid adding too much halide before reaction is well under way, or uncontrollable boiling and even fires may result. Halide is then added at such a rate as to maintain steady boiling of the ether. If boiling is allowed to stop before the formation *and use* of the reagent is completed, then an atmosphere of dry oxygen-free nitrogen should be provided as Grignard reagents are very rapidly hydrolysed and oxidized.

The induction period at the beginning of the preparation of a Grignard reagent increases rapidly with the water content. Magnesium bromide and iodide have a powerful drying effect, forming hydrates, and this, together with the attack on the metal surface, is why Grignard reactions may often be started by adding iodine (forming MgI_2) or ethylene dibromide (forming, mainly, $C_2H_4 + MgBr_2$).

Though diethyl ether is the preferred solvent by a wide margin, other solvents may be more appropriate in these circumstances:

(*a*) when high reaction temperatures are necessary,

(*b*) when the reaction product has a volatility comparable with that of ether,

(*c*) when the use of more basic ethers is necessary to promote Grignard reagent formation from relatively unreactive halides.

In the first case, the reagent may be prepared in ether in the usual way, the next reactant being added in a higher boiling solvent such as benzene, xylene or di-*n*-butyl ether. The mixture may then be heated well above 35°, when much ether (but not all of it) distils away.

The second case requires the complete absence of diethyl ether. The preparation of trimethylarsine (bp 52°),

$$AsCl_3 + 3MeMgI \longrightarrow Me_3As + 3MgClI$$

is an example in which a troublesome separation of Me_3As from Et_2O can be avoided by the use of di-*n*-butyl ether (bp 142°).

The third case is the most important, since many Grignard reagents which cannot be prepared (or only with difficulty) in diethyl ether can readily be prepared in the more basic tetrahydrofuran. The formation of phenylmagnesium chloride from chlorobenzene (*much* cheaper than PhBr) and magnesium in THF has already been mentioned, and it should be noted that an Et_2O : THF or even hydrocarbon : THF mixture commonly works well provided the THF : Mg molar ratio is at least unity. The use of the very basic ether 1,2-dimethoxyethane is complicated by the sparing solubility of $(MeOC_2H_4OMe)_2MgX_2$.

One of the best known and most valuable applications of THF has been in the preparation of vinyl and substituted vinyl Grignard reagents, developed to a large extent by H. Normant.

Organomagnesium halides may also be prepared in hydrocarbons, though careful attention to experimental detail is necessary to obtain good yields. This interesting development is due to D. Bryce-Smith. The hydrocarbon-soluble product from chlorobenzene and magnesium in tetrahydronaphthalene has the composition $(Ph_3Mg_2Cl)_x$.

It is often difficult to obtain Grignard reagents from polyhalides if there is scope for easy formation of magnesium halide, e.g.

$$C_3F_7MgI \longrightarrow C_3F_6 + MgIF$$

However, both bromopentafluorobenzene and hexachlorobenzene (the latter in THF) give the synthetically useful reagents C_6F_5MgBr and C_6Cl_5MgCl.

Considerable effort has been devoted to studies on the constitution of Grignard reagents, both in solution and in the crystalline state. The more reliable molecular weight determinations show an approach towards a monomeric constitution with increasing dilution in diethyl ether: such solutions must predominantly contain $RMgX(Et_2O)_n$ units, n probably being two. Ethereal solutions contain species whose average degree of association commonly exceeds two as the concentration rises to the 1–2 molar region, and phenylmagnesium bromide is approximately trimeric on average (just over 1 molar). This association could take place through halogen atoms leading to dimers such as 3.4 or to more associated species such as 3.5. Association through halogen is much more likely than through electron-deficient alkyl bridges since the dialkyls are considerably *less* associated than Grignard reagents at similar concentrations.

3.4   3.5

The structure 3.4 is indicated for *t*-butylmagnesium chloride because chloride has relatively good (and *t*-butyl relatively poor) bridging properties, because the degree of association does not appreciably exceed two even at high concentration, and since $(Bu^tMgClOEt_2)_2$ is a dimer in benzene solution (so is $[Et(Et_3N)MgBr]_2$ which consists of bromide-bridged dimers in the crystalline state).

Grignard reagents are monomeric in tetrahydrofuran over a wide concentration range. This difference from ether is due to the more strongly basic THF being able to displace halogen from co-ordination positions about the metal, better than the weaker base Et_2O with which halogen can compete with some success. There is evidence that solutions of ethylmagnesium chloride in THF contain Et_2Mg and $MgCl_2$ as well as $EtMgCl$.

The rapid precipitation of magnesium halide complexes when bases such as dioxan are added to Grignard reagents shows that exchange processes occur:

$$2RMgX + 2C_4H_8O_2 \longrightarrow R_2Mg + MgX_2(C_4H_8O_2)_2$$

This reaction provides a useful method for the preparation of ether solutions of R_2Mg. A study of the proton magnetic resonance spectra of various alkylmagnesium systems has shown not only that alkyl exchange takes place rapidly, but also that none of the species $RMgX$, R_2Mg, MgX_2 or complexes between these can have any prolonged existence in solution. It has also been found that exchange of pentafluorophenyl groups is much slower than that of alkyl or other aryl groups, and the ^{19}F resonances due to $(C_6F_5)_2Mg$ and C_6F_5Mg units can be distinguished in ethereal solutions of C_6F_5MgBr. As the solution is heated from room temperature to 94° the resonances due to *para* fluorine atoms coalesce to a single triplet. Resonances due to Me_2Mg and $MeMgBr$ can be distinguished in ether solutions of methylmagnesium bromide at low temperature (below $-100°$), but the equilibrium is very much in favour of $MeMgBr$.

Solutions of Grignard reagents in ether are electrically conducting and both anions and cations contain magnesium. The extent of ionization is small, and in a solvent of low dielectric constant such as ether there would be extensive formation of ion-pairs or higher aggregates. The organic products of electrolysis are consistent with the formation of radicals $R\cdot$ at the electrodes, since they consist of substances resulting from (a) hydrogen abstraction from solvent giving RH, (b) disproportionation to olefin and RH, and (c) coupling to R_2. Due to the relatively high stability of benzyl radicals, electrolysis of benzylmagnesium chloride gives $PhCH_2CH_2Ph$, whereas that of ethylmagnesium bromide gives mainly ethane and ethylene. Electrodes of zinc and lead are attacked giving zinc and lead alkyls, this being the basis for one process for making Me_4Pb and Et_4Pb for antiknock purposes.

To summarize, dilute (ca. less than 0·1 molar) solutions of Grignard reagents in diethyl ether consist mainly of solvated monomeric $RMgX$, which overall is the best as well as the simplest formula, together with very small amounts of ionic species. Dimers and more associated species, probably associated through the halogen, are increasingly abundant at higher concentrations.

The structures of crystalline $EtMgBr(Et_2O)_2$ and $PhMgBr(OEt_2)_2$ have been determined by X-ray diffraction. Both compounds consist of discrete monomer units in which the organic group, bromine, and two ether molecules are arranged tetrahedrally about the magnesium.

Though the products of reactions between Grignard reagents and other compounds are mostly intelligible on the basis of attack of the nucleophilic carbon on the most positive part of the reactant, e.g. the carbon atom of a carbonyl group, the detailed elucidation of reaction mechanisms has proved very difficult. Reactions with carbonyl compounds only are considered here. The first step is believed to be the reversible displacement of a solvent (normally Et_2O) molecule by the ketone

$$R_2C{=}O + R'MgX(OEt_2)_2 \ \rightleftharpoons \ R_2C{=}O \rightarrow MgR'X(OEt_2) + Et_2O$$

followed by an irreversible reaction of this complex with another molecule of Grignard reagent:

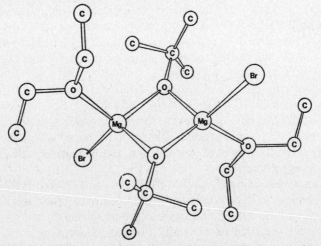

The final step is reaction between alkylmagnesium alkoxide and magnesium halide, regenerating Grignard reagent:

$$R_2R'C{\cdot}O{\cdot}MgR' + MgX_2 \ \rightleftharpoons \ R_2R'C{\cdot}O{\cdot}MgX + R'MgX$$

This mechanism, which has been tested in some detail by E. C. Ashby, accounts for a wide range of observations, including the drop in observed rate when half of the Grignard reagent has been consumed.

    The final product of the reaction between methylmagnesium bromide and acetone in ether is t-butoxymagnesium bromide, which crystallises as the dimeric ether complex whose structure (due to H.M.M. Shearer and P. T. Moseley) is shown in Figure 15.

Figure 15. Structure of t-butoxymagnesium bromide ether complex.

One interesting feature of this structure is that the three atoms bound both to alkoxy and to ether oxygen lie in one plane, as if there were O—Mg π-bonding, rather than pyramidally as normally expected for 3-covalent oxygen. This compound, which is dimeric both in benzene and in ether, is also obtained by the reactions

$$MeMgOBu^t + MgBr_2 \xrightarrow{Et_2O} MeMgBr + 1/2[Bu^tOMgBr, OEt_2]_2$$

and

$$Mg(OBu^t)_2 + MgBr_2 \xrightarrow{Et_2O} [Bu^tOMgBr, OEt_2]_2$$

Ether-free Bu^tOMgBr is insoluble in hydrocarbons and is likely to be polymeric.

Reactions between Grignard reagents and both ketones and cyanides go faster in more weakly basic, and slower in more strongly basic solvents. This is consistent with one step being the displacement of a solvent molecule by the ketone or cyanide. However, when the reaction may be regarded as involving the displacement, not of a solvent molecule but of some other species, then an increase in basicity of the solvent should *increase* reaction rate since it should increase the carbanionic character (and hence the reactivity) of the carbon atom bound to magnesium. Thus Et_2TlCl is converted to Et_3Tl in good yield by $EtMgBr$ in THF but not in diethyl ether. Here the carbanionic ethyl is displacing Cl^- from the Et_2Tl^+ cation, and the more carbanionic the ethyl the easier will the chloride be displaced.

(b) *Dialkyls and diaryls.* Though these can be obtained from magnesium and R_2Hg, this method is rarely used unless it is essential to have a product that must be quite free from halogen. The reaction needs particular care if carried out on more than about a 20 millimole scale, as it is commonly slow to start and then goes suddenly and exothermically.

For most purposes the dioxan precipitation method is suitable and convenient, using ether as solvent.

$$2RMgBr + 2C_4H_8O_2 \longrightarrow R_2Mg + MgBr_2(C_4H_8O_2)_2$$

It is advisable to add only a little more than the theoretical amount of pure dioxan *very slowly* to the stirred Grignard reagent, and then to continue stirring overnight.

Like Me_2Mg and Et_2Mg (see p 43), the known magnesium diaryls and the hydrocarbon-insoluble Pr^n_2Mg are electron-deficient polymers. Di-iso-butylmagnesium, which can be obtained from Bu^iMgCl and Bu^iLi in ether (in which LiCl is very sparingly soluble), followed by removal of ether by co-distillation with benzene, is only slightly soluble in benzene (ca. 0·1 moles per litre). Di-*sec*-butylmagnesium, in which there is chain branching α to magnesium and in which there is steric hindrance to polymerization though not to dimerization, is quite soluble in benzene. Di-*n*-amylmagnesium appears also to be soluble giving viscous solutions when concentrated; it is dimeric in dilute solution.

Biscyclopentadienylmagnesium can be made in good yield by a quite unusual preparative method for organometallic compounds, namely reaction between magnesium at 500–600° (it is still solid but has appreciable vapour pressure at these temperatures) and cyclopentadiene. It sublimes readily at 100° and is soluble in hydrocarbons as well as in ether. Its solubility in hydrocarbons provides no evidence against an ionic constitution, as salts tend to be insoluble in hydrocarbons only when the cation of one molecular unit can interact strongly with the anion of another (as in NaCl). In $(C_5H_5)_2Mg$, the Mg^{2+} cations (being sandwiched between two $C_5H_5^-$ rings) would not be able to interact strongly with $C_5H_5^-$ rings of neighbouring $(C_5H_5)_2Mg$ molecules (or ion-triplets). Intermolecular forces would therefore not be strong, and this would allow solubility in hydrocarbons.

(*iii*) *Beryllium*. Small amounts of Me_2Be can conveniently be made by the dialkylmercury method $(Be + Me_2Hg)$, which has, for example, been applied to the preparation of $(CD_3)_2Be$ for spectroscopic purposes. Diphenylberyllium is also obtained from Ph_2Hg and beryllium.

The dialkyls are normally made from beryllium chloride and the Grignard reagent. After removal of much of the magnesium halide, the dialkyls (ethyl to the butyls) are distilled under reduced pressure as ether complexes, from which ether can be separated completely by prolonged refluxing at low pressure. As mentioned in an earlier section, dimethylberyllium is polymeric and most other dialkyls investigated are dimeric. Di-*t*-butylberyllium, in contrast, is monomeric and is the most volatile known beryllium compound (in respect of vapour pressure at room temperature). It cannot therefore be separated from ether by refluxing the ether complex under reduced pressure, and the ether-free material is obtained by reaction with a stronger Lewis acid:

$$Bu^t_2BeOEt_2 + 1/2BeCl_2 \longrightarrow Bu^t_2Be + 1/2BeCl_2(OEt_2)_2$$

The dimeric (in benzene) Grignard analogue 3.6 is a by-product; it is monomeric in dilute solution in ether.

3.6

Di-*t*-butylberyllium and its ether complex are of interest because their pyrolysis (up to ca. 200°) gives the polymeric beryllium hydride.

$$Bu^t_2BeOEt_2 \longrightarrow BeH_2 + 2C_4H_8 + Et_2O$$

Cautious pyrolysis of di-isobutylberyllium gives the oligomeric (in benzene) Bu^iBeH which adds to olefins very much faster than do the alkylberyllium hydride ether complexes. Alkylberyllium hydrides, even as ether complexes, react very rapidly with azomethine and carbonyl bonds, giving amino- and alkoxy-beryllium alkyls (discussed below pp 113-117), e.g.

$$4MeBeH + 4PhCHO \longrightarrow (MeBeOCH_2Ph)_4$$

Alkylberyllium hydrides can be obtained in ether by the reaction,

$$RBeBr + LiH \longrightarrow RBeH + LiBr$$

and can be isolated as ether or amine complexes which are dimeric in benzene. These hydrides contain an electron-deficient BeH_2Be bridge analogous to the BH_2B bridge in diborane, but differ sharply in that the BeH_2Be bridge as in 3.7 is not split by reaction with bases such as tri-methylamine.

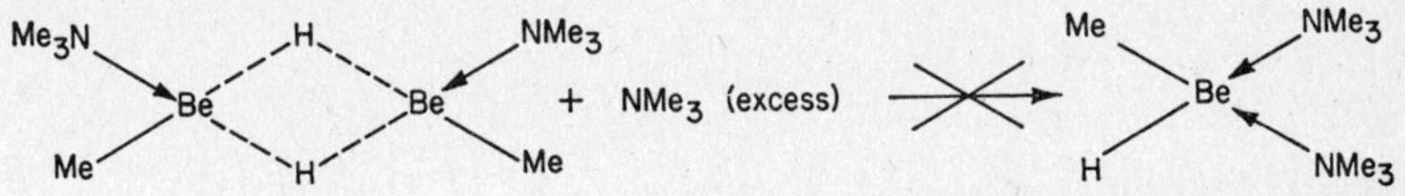

3.7

Though no evidence has been obtained for *cis-trans* isomerism about the more labile BH_2B bridge, the temperature dependence of the proton magnetic resonance spectrum of 3.7 (Figure 16) indicates such isomers and has given the main thermodynamic parameters (note the relatively large entropy difference due to the greater ordering of solvent molecules by the highly polar *cis* than by the nonpolar *trans* form).

Diethylberyllium forms an anionic complex, $NaOEt_2[Et_2BeH]$, when stirred with sodium hydride in ether, and part of its crystal structure is shown in Figure 17. The presence of the $Et_4Be_2H_2$ unit should be noted, the $[Et_4Be_2H_2]^{2-}$ anion being isoelectronic with $Et_4B_2H_2$. It is likely that the bridging hydrogen atoms (located in the structure analysis) are electronically more closely associated with beryllium than with sodium. The four rather than two thin lines from oxygen indicate that the ethyl groups of the coordinated ether molecules can occupy two alternative positions.

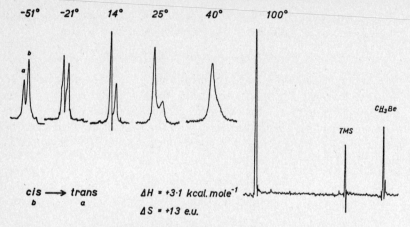

Figure 16. $(Me_3N.BeMeH)_2$ in C_7D_8. Effect of temperature on CH_3N resonance.

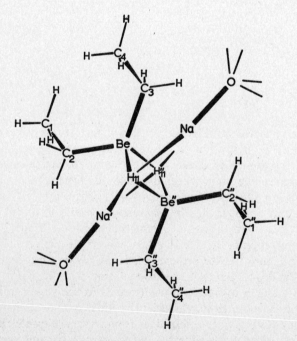

Figure 17. Structure of $[NaOEt_2]_2[Et_4Be_2H_2]$. (After G. W. Adamson and H. M. M. Shearer, *Chem. Comm.*, 1965, 240). Ether ethyl groups are omitted.

The Group IIB Metals

(i) *Derivatives of the type RMX.* These have a curious history. Ethylzinc iodide was first described by Frankland (1849) who made it from ethyl iodide and zinc, at a stage in the development of chemistry at which the ethyl group was represented C_4H_5 on account of uncertainties about atomic weights. Though diethylzinc, and other dialkyls R_2Zn, obtained by the thermal disproportionation of RZnI were valuable synthetic reagents until they were almost entirely superseded by Grignard reagents (from about 1900), the nature of RZnX long remained obscure. During the last ten years many chemists have been persuaded that alkylzinc halides did not exist as such; they were to be represented $R_2Zn.ZnX_2$. The structure of ethylzinc iodide (1966) is shown in Figure 18, which

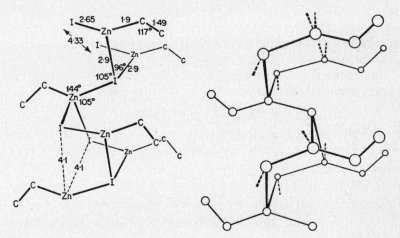

Figure 18. Crystal structure of the ethylzinc iodide polymer. (After P. T. Moseley and H. M. M. Shearer, *Chem. Comm.*, 1966, 876).

indicates that each zinc is 4-co-ordinate (bound to one carbon and three iodines) and each iodine is bound to three zinc atoms.

Surprisingly little is known about alkyl or aryl cadmium halides, and at the time of writing nothing is known of their structures.

In contrast, alkyl and aryl mercury halides (and numerous other derivatives in which X is a relatively electronegative group other than halogen) have been known for a long time, and are monomeric linear molecules.

The alkylzinc iodides are formed from alkyl iodides and zinc which is activated by copper (e.g. as zinc-copper couple, or alloy, or simply mixing the metal powders). In basic solvents such as dimethylformamide and 1,2-dimethoxyethane, the lower alkyl bromides also react with metallic zinc.

The lower alkylzinc halides ($X = Cl$, Br, I) can be obtained from the dialkyls:

$$Et_2Zn \begin{cases} + ZnCl_2 \longrightarrow 1/2[EtZnCl]_4 \text{ mp } 68° \\ + ZnBr_2 \longrightarrow 1/2[EtZnBr]_4 \text{ mp } 81° \\ + ZnI_2 \longrightarrow 2/n[EtZnI]_n \text{ (polymer)} \end{cases}$$

Organozinc bromides are formed as intermediates in the *Reformatsky* reaction in which a zinc derivative of an α-bromo-ester reacts with an aldehyde or ketone:

$$BrCH_2CO_2Et \xrightarrow{Zn} [BrZnCH_2CO_2Et] \xrightarrow[\text{(ii) } H_2O]{\text{(i) } R'_2CO} R'_2COH \cdot CH_2CO_2Et$$
$$\text{not isolated}$$

The above intermediate has been prepared in high yield using methylal as solvent, and then allowed to react with carbonyl compounds giving the usual products.

One more compound belonging to the class RZnI should be mentioned, namely iodomethylzinc iodide, formed in ether solution:

$$CH_2I_2 + Zn \longrightarrow ICH_2ZnI$$

If this reaction is carried out in the presence of olefins, then cyclopropanes are obtained, often in good yield, by a methylene transfer process (the Simmons-Smith reaction). In the absence of olefin, ethylene is the main product:

$$ICH_2ZnI + olefin \longrightarrow cyclopropane + ZnI_2$$

The reaction between olefin, CH_2I_2 and Et_2Zn also gives cyclopropanes, but is faster.

The alkyl and aryl mercury halides are accessible by a wide variety of routes, several of which have already been mentioned (Chapter 2). Generally, the Grignard method is quite convenient:

$$RMgX + HgCl_2 \longrightarrow RHgX + MgCl_2$$

If $X = Br$ or I, then the product is mainly RHgBr or RHgI. If aluminium alkyls or the sesquihalides are available, then alkylation of $HgCl_2$ by these in a solvent such as CH_2Cl_2 is convenient and can be carried out on a fairly big scale:

$$Me_3Al_2Cl_3 + 3HgCl_2 \longrightarrow 3MeHgCl + 2AlCl_3$$

The reaction between mercury and methyl iodide in sunlight was discovered in 1853 (oddly enough, in England) by Frankland: it is still a very simple and effective way of making methylmercury iodide. The reaction between mercury and allyl iodide goes with remarkable facility when the reagents are gently warmed together. The product is believed to contain a π-bond between olefin and mercury as well as a σ-bond between methylene and mercury, though allylmercury systems are labile as shown by their n.m.r. spectra.

The formation of organomercury compounds by the use of halides and sodium amalgam, diazo compounds, diazonium salts, decarboxylation reactions, mercuration of aromatic compounds, and reactions between mercury halides (or acetates) and olefins or acetylenes has been discussed in Chapter 2 and is not further elaborated here.

The electrolysis of solutions of the lower alkylmercury halides in liquid ammonia results in the deposition on the cathode of an electrically conducting solid of metallic appearance, which decomposes at room temperature into equimolar proportions of R_2Hg+Hg. The substance is easiest to obtain when $R = CH_3$, and appears to be an *organic metal* consisting of CH_3Hg cations and an equivalent number of free electrons. Its possible formulation as consisting of free radicals, or $RHgHgR$, or amalgams of $Hg+RHg\cdot$, has been excluded in various ways. The electrolysis of water-pyridine solutions of MeHgOAc at room temperature leads directly to dimethylmercury.

$$MeHg^+ + e^- \longrightarrow MeHg\cdot \longrightarrow 1/2Hg + 1/2Me_2Hg$$

(ii) *Dialkyls and diaryls.* These are, in general, conveniently obtained by the action of Grignard reagents on anhydrous metal halides, e.g. extracting $HgCl_2$ (Soxhlet) into boiling methylmagnesium bromide. The mercury compounds R_2Hg are much easier to separate from the reaction mixture, since unlike R_2Zn and R_2Cd, they normally resist hydrolysis.

Diphenylmercury is easily made by reduction of almost any phenyl-mercury compound PhHgX (some of which are commercially available and fairly cheap) by hydrazine hydrate and sodium carbonate in aqueous ethanol.

Zinc alkyls, R_2Zn, can be made by the Grignard method but, after separation of solid magnesium halides, have to be isolated by distillation under reduced pressure. Clearly, one would not prepare Me_2Zn (bp 44°) using a Grignard reagent in diethyl ether (bp 35°), but Me_2Zn, Et_2Zn and Bu^i_2Zn are very easily made from the corresponding aluminium alkyls, all

of which are commercially available. Zinc chloride or the more easily
dried acetate can be used: no solvent is needed.

$$2Et_3Al + Zn(OAc)_2 \longrightarrow 2Et_2AlOAc + Et_2Zn$$

Unsymmetrical zinc dialkyls can be prepared, e.g.

$$EtZnI + Pr^nMgBr \longrightarrow EtPr^nZn + MgBrI$$

but have to be isolated at temperatures not much exceeding room temperature. They slowly disproportionate, e.g.

$$2EtPr^nZn \rightleftharpoons Et_2Zn + Pr^n_2Zn$$

probably through electron-deficient alkyl-bridged dimeric intermediates.
Alkyl exchange is facilitated when some base is present. Similarly, unsymmetrical RHgR' can be made: these undergo exchange giving a near
statistical distribution of R_2Hg, R'_2Hg and RR'Hg when the groups R and
R' are fairly similar, e.g. both aryl or saturated alkyl. Otherwise redistribution can be far from statistical, e.g.

$$(C_6F_5)_2Hg + Me_2Hg \longrightarrow 2C_6F_5HgMe$$

The two metal-carbon bonds in R_2Zn, R_2Cd and R_2Hg are collinear
but the change of bond length (in the dimethyls) descending the subgroup
is worth noting: Zn—C, 1·929 Å; Cd—C, 2·112 Å; Hg—C, 2·094 Å,
all ±0·004–5 Å. The anomalous position of cadmium is reflected also
in the boiling point sequence, Zn, 44°; Cd, 106°; Hg, 92°, and is attributed
to bigger van der Waals forces between Me_2Cd than between Me_2Hg
molecules.

Though R_2Hg is thermodynamically unstable with respect to decomposition to mercury and hydrocarbons, dimethylmercury is stable indefinitely at room temperature. The other dialkyls decompose, sometimes
slowly, depositing mercury. The diaryls are usually stable indefinitely at
room temperature. Those whose crystal structure has been determined all
contain C—Hg—C bonds which are collinear or nearly so, an example
being shown in Figure 19.

Biscyclopentadienylmercury, easily made by the reaction

$$2C_5H_5Tl + HgCl_2 \longrightarrow 2TlCl + (C_5H_5)_2Hg$$

is an interesting compound. Its infrared spectrum and its reaction with
maleic anhydride show that it has a σ-bonded constitution, but its proton
magnetic resonance spectrum consists of a single resonance at room
temperature. At low temperatures the p.m.r. spectrum becomes quite
complex: evidently some process takes place at room temperature which
has the effect of making all the protons magnetically equivalent over
periods of the order of 0·01–0·001 seconds. Clearly the σ-bond must

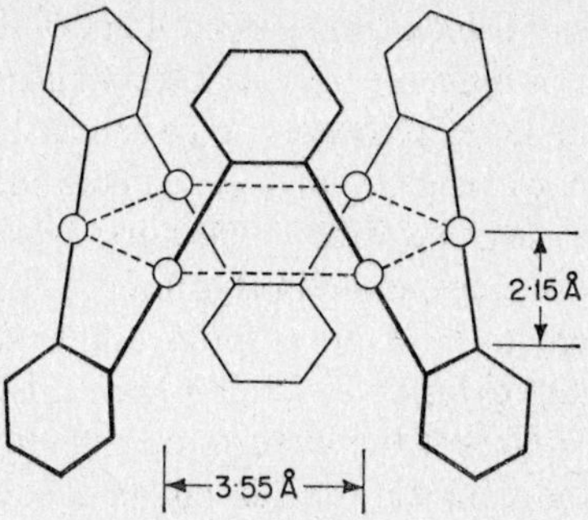

Figure 19. Structure of o-phenylenemercury. (After D. Grdenic, *Chem. Ber.*, **92**, 1959, 231).

shift from one carbon atom to another, and this is believed to take place through a transition state in which one of the double bonds is held to the mercury by a π-bond:

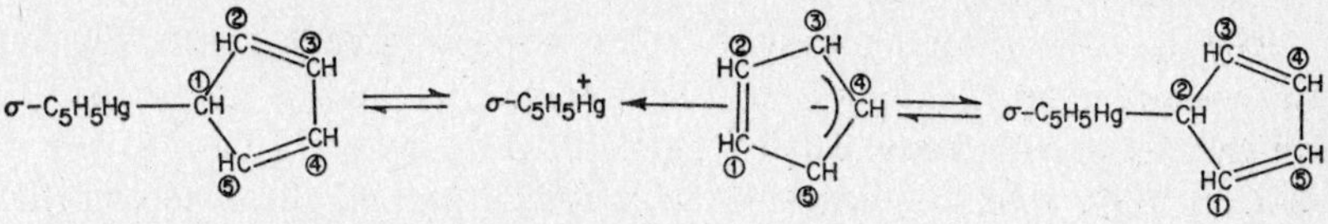

Many examples of this kind of behaviour have been found.

One more class of organomercury compound deserves special mention, namely the trihalomethyls, whose chemistry has largely been developed by D. Seyferth. For example, the reaction

$$PhHgCl + CHCl_2Br + KOBu^t \longrightarrow PhHgCCl_2Br + KCl + Bu^tOH$$

which involves the nucleophilic displacement of Cl by CCl_2Br^-, gives a product which very easily transfers CCl_2 groups to olefins either as dichlorocarbene or in a concerted bimolecular process, the bromine remaining bound to mercury:

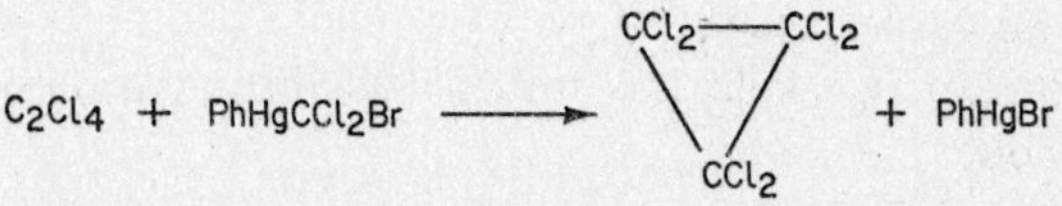

An advantage of this method of obtaining cyclopropanes from trihalo-methyl compounds is that basic conditions are not needed.

Organocadmium compounds have found synthetic use for *ketone synthesis*:

$$2RCOCl + R'_2Cd \longrightarrow 2RCOR' + CdCl_2$$

F

This preparative method is more convenient than the direct use of Grignard reagents since these combine not only with acid halides but also with the ketones produced. The latter reaction can be reduced by reverse addition, i.e. adding the Grignard reagent to the acid chloride, but this is apt to cause some bother to those not used to handling air-sensitive compounds. The normal procedure using cadmium is to prepare the Grignard solution in the usual way then, without interrupting the boiling of the ether, to add powdered cadmium chloride which has been dried to constant weight at 110°. Removal of much of the ether and replacement by dry benzene before addition of the acid chloride is advantageous since the higher boiling temperatures facilitate rapid reaction, and side reactions due to ester formation (from acid chloride and ether) and to enolization of the ketone are reduced.

This method allows the preparation of ketones from acid chlorides containing other functional groups such as CO, CO_2R, and CN, though some of these react slowly with organocadmium compounds under the conditions normally used.

The reactivity of organocadmium compounds is greatly affected by the presence of magnesium (and other soluble) halides, and, of course, nearly all the ketone syntheses carried out with these reagents have involved $R_2Cd + MgX_2$ mixtures, possibly with $RCdX$ as well. Benzoyl chloride reacts only very slowly with R_2Cd ($R = Me$, Et, Bu^n or Ph), but gives acetophenone in good yield with $Me_2Cd + LiBr$. For another example, the reaction between Et_2Cd and $MeCOCl$ goes about 20% to completion in 20 minutes at room temperature in benzene, but is 97% complete under the same conditions except for the addition of 2 mols. $MgBr_2$-ether complex (per mole of Et_2Cd). The activating effect of magnesium halides is in the order $MgI_2 > MgBr_2 > MgCl_2$. The reason for these effects is not clear, though in some cases the metal halides are believed to facilitate the formation of reactive acylonium intermediates from acid chlorides.

The low reactivity of R_2Cd in the absence of soluble halide provides an example of a well-known synthetic method of long standing proving more complex than earlier suspected.

Pure cadmium dialkyls are rather unstable and prone to thermal decomposition.

The Group III elements

(i) Boron

(a) *Preparation of organoboron compounds using other organometallics.*
Boron, the only non-metal among the elements of the first three groups of the periodic table, differs from the others in that the element itself is too unreactive (and anyway much too expensive) for its action on organic

halides or on other organometallic compounds to be used as preparative routes to organoboranes. The most generally useful route to its trialkyls or triaryls is that between another organometallic compound (preferably RMgX or R_3Al) and a boron halide (preferably BF_3 as an ether complex, e.g. BF_3, OEt_2 – boron trichloride tends to cleave ethers) or a borate ester such as $B(OEt)_3$ or $B(OMe)_3$:

$$Et_2O,BF_3 + 3RMgX \longrightarrow R_3B + 3MgXF + Et_2O$$

The volatility of the lighter alkyls of boron (Me_3B, bp $-22°$; Et_3B, bp $95°$) and the weakness of their interaction with ethers allows them to be distilled directly from the reaction mixture. Higher alkyls and aryls can be separated after hydrolysis of the mixture, as organoboranes are unaffected by water at normal temperatures, although exclusion of oxygen is essential. In the aryl series, it is important to avoid an excess of Grignard reagent which results in the formation of BAr_4^- anions, products of a type which are readily formed in both aryl and alkyl series when organolithium compounds are used as the arylating or alkylating agents.

The most satisfactory alkylating agents are aluminium alkyls (when commercially available). No quaternary salts are formed, and generally no solvent or diluent is required. For example, a mixture of triethylaluminium and ethyl orthoborate warms spontaneously to over $100°$, and triethylborane can be distilled off in over 90% yield:

$$B(OEt)_3 + Et_3Al \longrightarrow Et_3B + Al(OEt)_3$$

Organoboron compounds with only one organic group attached to boron can be prepared by use of a suitably limited amount of Grignard or organo-aluminium compound,

$$BF_3,OEt_2 \xrightarrow[\text{(ii) hydrolysis}]{\text{(i) PhMgBr in } Et_2O} PhB(OH)_2$$

or alternatively by use of organo-tin or -mercury compounds, which readily replace one halogen of a boron halide by an organic group. More forcing conditions with these last reagents afford diorganoboron compounds R_2BX,

$$Ph_4Sn + 4BCl_3 \longrightarrow 4PhBCl_2 + SnCl_4$$

$$Ph_4Sn + 4PhBCl_2 \longrightarrow 4Ph_2BCl + SnCl_4$$

which otherwise are accessible by experimentally rather inconvenient exchange reactions ($2R_3B + BX_3$) or by cleavage of one organic group from a triorganoborane by chelating acids such as acetylacetone (see p 88).

(b) *Addition of B—H bonds to olefins or acetylenes: hydroboration.* In the presence of ethers and at or near room temperature, compounds containing B—H bonds react with olefins or acetylenes to form organoboron compounds:

$$RCH{:}CH_2 + HB\diagdown^{\diagup} \longrightarrow RCH_2CH_2B\diagdown^{\diagup}$$

$$RC{:}CH + HB\diagdown^{\diagup} \longrightarrow RCH{:}CHB\diagdown^{\diagup}$$

Although clearly of use in the synthesis of certain alkylboron compounds, particularly tri-*n*-alkylboranes and alkylboron hydrides, the hydroboration reaction (developed mainly by H. C. Brown) is of considerable significance in synthetic organic chemistry, as the resulting organoboron compound can be used (without being isolated) as an intermediate in the conversion of olefins or acetylenes into other hydrocarbons (paraffins or olefins), alcohols, aldehydes, ketones, carboxylic acids or amines.

A convenient procedure is to bubble diborane in a stream of nitrogen through a solution of the olefin in an appropriate ether (commonly diglyme or tetrahydrofuran). Reaction is rapid at room temperature, and excess diborane is conveniently removed by washing exit gases with acetone. Tri-*n*-hexylborane can be prepared in $> 90\%$ yield in this way:

$$6Bu^nCH{:}CH_2 + B_2H_6 \longrightarrow 2(C_6H_{13})_3B$$

Alternatively, diborane may be generated *in situ* by adding boron trifluoride-ether complex to a solution of the olefin and sodium borohydride in diglyme:

$$12RCH{:}CH_2 + 3NaBH_4 + 4BF_3{,}OEt_2 \longrightarrow 4(RCH_2CH_2)_3B + 3NaBF_4 + 4Et_2O$$

Although other procedures may be useful in particular instances, the essential ingredients are always a *hydride,* which need not contain boron (e.g. $LiBH_4$, $NaBH_4$, KBH_4, $LiAlH_4$, NaH, $py{,}BH_3$ or $Me_3N{-}BH_3$); an *acid,* which should contain boron if the hydride does not (e.g. BF_3, BCl_3, $AlCl_3$, $AlCl_3 + B(OMe)_3$, $TiCl_4$, HCl or H_2SO_4); and normally a suitable *ethereal solvent,* usually chosen so that its boiling point or miscibility with water allows easy separation of the product.

The direction of addition of B—H bonds to olefins is such as to attach the boron generally to the less substituted of the two carbon atoms of the double bond, as the orientation is governed by the polarities of the olefin and of the $B^{\delta+}-H^{\delta-}$ bond. Studies with cyclic olefins and with acetylenes have shown *cis* addition of B—H to occur, and a four-centre transition state appears likely:

Readily polarized olefins may give appreciable proportions of both possible products:

$$\mathrm{PhCH{:}CH_2} \xrightarrow{\mathrm{HB}\langle} 20\%\ \mathrm{PhCHMe{\cdot}B}\langle + 80\%\ \mathrm{PhCH_2CH_2B}\langle$$

However, in such cases attachment of boron to the less substituted carbon can be ensured by hydroboration with a substituted diborane, particularly a tetra-alkyldiborane R_2BHBHR_2 in which the groups R are bulky. Di-*sec*-isoamylborane ('di-siamylborane') $(Me_2CHCHMe)_2BH$, itself prepared from B_2H_6 and $Me_2C{:}CHMe$,

$$[\mathrm{BH_3}] \xrightarrow[\text{fast}]{\text{Excess } \mathrm{Me_2C{:}CHMe}} [(\mathrm{Me_2CHCHMe})_2\mathrm{BH}] \xrightarrow[\text{very slow}]{\text{Excess } \mathrm{Me_2C{:}CHMe}} (\mathrm{Me_2CHCHMe})_3\mathrm{B}$$

is one such reagent which has found wide application. With styrene for example it gives only 2% of the α-boryl derivative (contrast 20% with diborane). Others are t-hexylborane $Me_2CHCMe_2BH_2$ (from B_2H_6 and $Me_2C{:}CMe_2$) and 9-borabicyclo[3.3.1]-nonane ('9BBN', prepared by the hydroboration of cyclo-octa-1,5-diene):

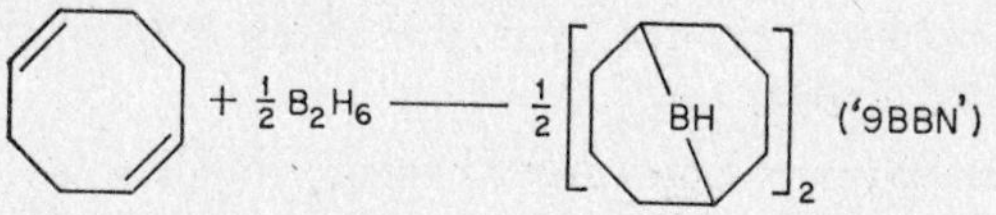

A further important aspect of the hydroboration reaction is its reversibility. Alkylboranes with the structural feature $>\mathrm{CH{-}\overset{\vee}{C}{-}B}<$ tend to lose olefin when heated strongly:

$$>\mathrm{CH{\cdot}C{\cdot}B}< \underset{25°;\ \text{ethers}}{\overset{150°\text{–}200°}{\rightleftharpoons}} >\mathrm{C{:}C}< + \mathrm{H{-}B}<$$

Successive addition/elimination reactions can therefore result in the isomerization of olefins, e.g.

$$\mathrm{MeCH{:}CHMe} \xrightleftharpoons{\rangle\mathrm{BH}} \mathrm{MeCH(B\langle)CH_2Me} \underset{\rangle\mathrm{BH}}{\overset{>180°}{\rightleftharpoons}} \mathrm{H_2C{:}CHCH_2Me} \xrightleftharpoons{\rangle\mathrm{BH}} \mathrm{H_2C(B\langle)CH_2CH_2Me}$$

Hydroboration of an internal olefin followed by thermal isomerization of the organoboron product accordingly gives an alkylborane in which boron is attached to a terminal carbon. The terminal olefin can then be isolated by displacement from the alkylborane using an involatile olefin:

$$\mathrm{Bu^n{}_3B} + 3\ \text{1-decene} \xrightarrow{100°} \mathrm{decyl_3B} + 3\mathrm{CH_3CH_2CH{=}CH_2}\uparrow$$

Because the overall change effected is conversion of an internal olefin into a terminal one, this process effects contra-thermodynamic isomerization of the internal olefin (usually, terminal olefins may be converted into internal olefins under the influence of acid catalysts).

Both boron and aluminium hydrides react with acetylenes by *cis*-addition, forming vinyl derivatives, which form *cis*-olefins when treated with acetic acid:

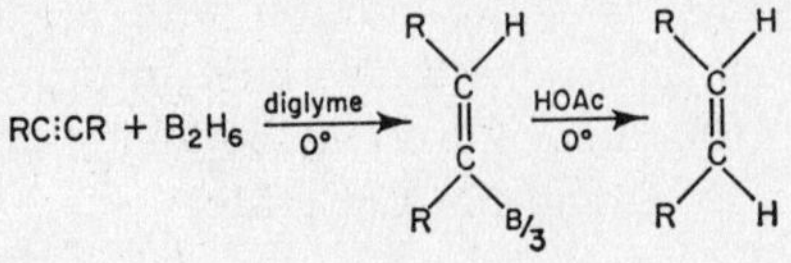

$$RC{:}CR + B_2H_6 \xrightarrow[0°]{diglyme} \quad \underset{R \quad B/_3}{\overset{R \quad H}{C{=}C}} \xrightarrow[0°]{HOAc} \quad \underset{R \quad H}{\overset{R \quad H}{C{=}C}}$$

Many further products are preparable from olefins *via* hydroboration reactions, as illustrated by the following reactions of alkylboranes.

(*i*) Alkaline hydrogen peroxide yields an *alcohol* (and boric acid), i.e. overall *cis anti-Markovnikov hydration* of the olefin.

(*ii*) Protonation by acetic or propionic acid yields a hydrocarbon, i.e. overall *hydrogenation*.

(*iii*) Oxidation with chromic acid gives *ketones* (from $RR'CHB\langle$ or *acids*, (from $RCH_2B\langle$).

(*iv*) Alkaline silver nitrate causes *coupling* of the alkyl groups.

(*v*) Treatment with Et_2NCl gives an alkyl chloride, i.e. overall *anti-Markovnikov addition of hydrogen chloride*.

(*vi*) Chloramine or hydroxylamine-O-sulphonic acid gives an *alkyl amine*.

(*vii*) Carbonylation at 100–125° affords *carbinols* R_3COH

$$R_3B \xrightarrow[\text{diglyme; 100–125°}]{\text{CO; 1 atm}} [R_3CBO] \xrightarrow[\text{NaOH}]{H_2O_2} R_3COH$$

(*viii*) Carbonylation in the presence of water affords *ketones*

$$RR'R''B \xrightarrow[\text{wet diglyme; 100°}]{\text{CO; 1 atm}} [RB(OH)CR'R''OH] \xrightarrow[\text{NaOH}]{H_2O_2} R'R''CO$$

(*ix*) Carbonylation in the presence of a borohydride effects *oxymethylation*:

$$R_3B + CO \xrightarrow[\text{or LiBH}_4/\text{THF; 45°}]{\text{NaBH}_4/\text{diglyme}} \xrightarrow{\text{KOH}} RCH_2OH$$

(*x*) Reaction with α-bromo ketones or esters under the influence of $KOBu^t$ effects substitution of alkyl for bromine, e.g.

$$R_3B + BrCH_2CO_2Et \xrightarrow{\text{Bu}^t\text{OK/Bu}^t\text{OH; 0°}} RCH_2CO_2Et$$

(*xi*) Reaction of R_3B with iodine in the presence of base affords iodides RI.

(*xii*) Reaction with α,β-unsaturated carbonyl compounds effects addition of RH to the double bond:

$$R_3B + CH_2:CHCOMe \rightarrow RCH_2CH:C(Me)OBR_2 \xrightarrow{H_2O} RCH_2CH_2COMe$$

(*c*) *Other ways of making and breaking B—C bonds.* The hydroboration reaction is not the only reaction by which insertion of olefins or acetylenes into B—X bonds gives rise to boron-carbon links, although it is by far the most important. Acetylenes react smoothly with boron trichloride or more vigorously with boron tribromide to form 2–halovinylboranes, e.g.

$$BCl_3 + CH:CH \longrightarrow 90\% \ Cl_2BCH:CHCl$$

Olefins react less readily with boron trichloride than with diborane, many being polymerized, although boron trichloride adds to norbornadiene. The insertion of methylene units into B—X bonds using diazomethane was discussed in Chapter 2.

Friedel-Crafts reactions involving boron halides have been used as key cyclization steps in the preparation of borazarenes, though yields are rarely high.

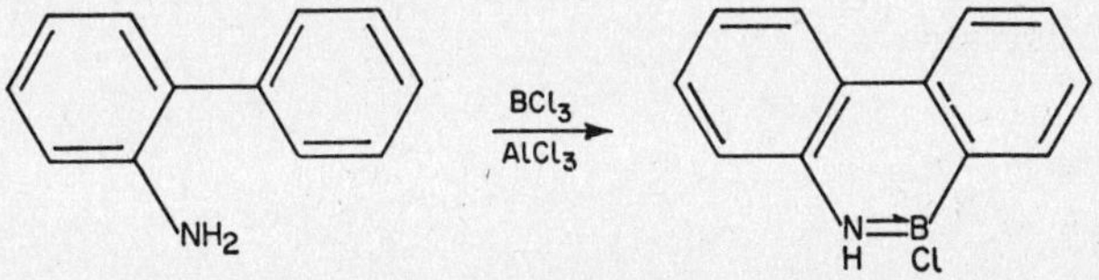

Although the trialkyls and triaryls of boron are unique among such derivatives of Group III elements in being undecomposed by water under normal conditions, a consequence of the low polarity of the B—C bond (see p 36), they are nevertheless very sensitive to oxygen. The lower (highly volatile) trialkyls are spontaneously inflammable and burn with a green-tinged flame. Controlled oxidation leads to borinic and boronic esters R_2BOR and $RB(OR)_2$ *via* peroxides R_2BOOR, themselves possibly formed *via* complexes R_3B,O_2 which however rearrange too rapidly to be isolated. Oxidation with anhydrous tertiary amine N-oxides (e.g. Me_3NO) provides a method for the quantitative determination of boron-carbon bonds, as the amine released can be determined acidimetrically:

$$R_3B + 3Me_3NO \longrightarrow B(OR)_3 + 3Me_3N \uparrow$$

The halogens react readily with trialkylboranes. The lower trialkyls inflame in chlorine or bromine, but tripropylborane and iodine at 150° react smoothly to form iododipropylborane, Pr_2BI. Chlorination of methyl groups attached to boron, without cleavage of the B—C bond, is possible using gaseous chlorine at $-96°$:

$$Me_3B + Cl_2 \xrightarrow{-96°} ClCH_2BMe_2 + HCl$$

As a route to organoboron halides R_2BX or RBX_2, which are useful intermediates in the preparation of other organoboron compounds, the direct halogenation of boron trialkyls or triaryls is both difficult to control and wasteful of organic groups. Reactions using hydrogen halides suffer similar disadvantages, although the butylboron chlorides can be prepared thus:

$$Bu^n{}_3B \xrightarrow[110°]{HCl} Bu^n{}_2BCl \xrightarrow[110°]{HCl/AlCl_3} Bu^nBCl_2$$

Exchange reactions between organoboranes R_3B and trihalides BX_3 afford alkylhaloboranes without loss of alkyl groups, although such reactions require temperatures generally $>150°$, and a mixture of mono- and dihalide normally results and has to be separated by low pressure distillation:

$$Et_3B + BCl_3 \underset{}{\overset{200°}{\rightleftharpoons}} Et_2BCl + EtCl_2$$

Another route to the dihalides involves boroxines:

$$R_3B + B_2O_3 \xrightarrow{200°} (RBO)_3 \xrightarrow{BCl_3} RBCl_2 + B_2O_3$$

These and other reactions of organoboranes are summarized in the following diagram:

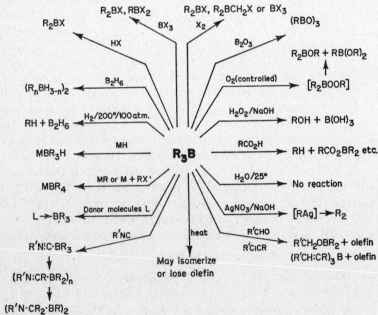

(*d*) *Unsaturated organoboron compounds.* Triarylboranes are relatively involatile solids which dissolve in inert solvents as monomers, and are normally prepared by the Grignard method. They are more resistant to oxidation than the simple trialkyls, and examples with bulky aryl groups such as 1-naphthyl or mesityl are sufficiently (kinetically) stable to oxygen to allow their isolation in air.

Triarylboranes in ether solution react with alkali metals to form paramagnetic anion radicals $Ar_3B^{\cdot-}$ isoelectronic with triarylmethyl neutral radicals, and the colours of the two series are similar. Association to diamagnetic anions $Ar_3BBAr_3^{2-}$ may occur provided that the bulk of the aryl groups is not too great and provided that co-ordination with the ether solvent does not occur.

Triarylborane-alkali metal compounds are very reactive; iodine in ether is immediately decolorized, and methanol gives Ar_3BH^-, e.g.

$$2Ph_3B^{\cdot-}Na^+ + MeOH \longrightarrow Na^+[Ph_3BOMe]^- + Na^+Ph_3BH^-$$

Triphenylborohydrides $MBPh_3H$, can also be prepared from ethereal Ph_3B and NaH or LiH, or from $LiH + Ph_3B$ at 180° with no solvent.

Tetraphenylborates $M^+BPh_4^-$ are obtained by analogous reactions, e.g.

$$PhLi + Ph_3B \xrightarrow{Et_2O} LiBPh_4$$

Lithium tetraphenylborate is stable to boiling water, in which it dissolves as the ions Li^+ and BPh_4^-, and is decomposed by acids only at 80° or more. Sodium tetraphenylborate (from BF_3 or $NaBF_4 + an$ excess of PhMgBr, followed by aqueous NaCl) is a valuable analytical reagent, as a precipitant for potassium, rubidium, caesium, thallium, various nitrogen bases, and a number of other cations, e.g. Ph_4P^+ and $Co(C_5H_5)_2^+$. The ammonia and methylamine salts decompose on heating (providing a route to Ph_3B):

$$NH_4BPh_4 \xrightarrow{240°} Ph_3B + C_6H_6 + NH_3$$

Vinylboranes are generally prepared from the appropriate vinyl derivative of another metal, although some can be prepared by the hydroboration of acetylenes:

$$12RC:CR' + 3NaBH_4 + 4BF_3 \longrightarrow 4(RCH:CR')_3B + 3NaBF_4$$

In vinylboranes, the vacant p orbital of the boron may be used in a delocalization of the π-electrons of the ethylenic bond, so that the structure may be written $CH_2\overset{\delta+}{=}CH\overset{\delta-}{=}BR_2$ rather than $CH_2=CH-BR_2$. Consequently, polar addition reactions of the vinyl groups tend to place the negative group on the β carbon:

$$CH_2{:}CHB(OBu)_2 + BH_3{\cdot}THF \longrightarrow \underset{\underset{BH_2}{|}}{MeCHB(OBu)_2} \xrightarrow[2.H_2O]{1.BuOH} MeCH\overset{B(OH)_2}{\underset{B(OH)_2}{}}$$

Polar or radical-initiated additions to the vinyl groups make vinylboranes of some importance as intermediates both in the preparation of other organofunctional boranes and in synthetic organic chemistry. Allylboranes $CH_2{:}CHCH_2BR_2$ and alkynylboranes $RC{:}CBR'_2$ are also relatively reactive.

Among boron-carbon ring systems, the unsaturated seven-membered borepin ring is interesting in that it is isoelectronic with the tropylium cation and has aromatic character. The benzoborepin 3.8 which in its chemical, u.v. and n.m.r. spectroscopic properties shows typical aromatic characteristics, can be prepared *via* the tin analogue (which is not aromatic):

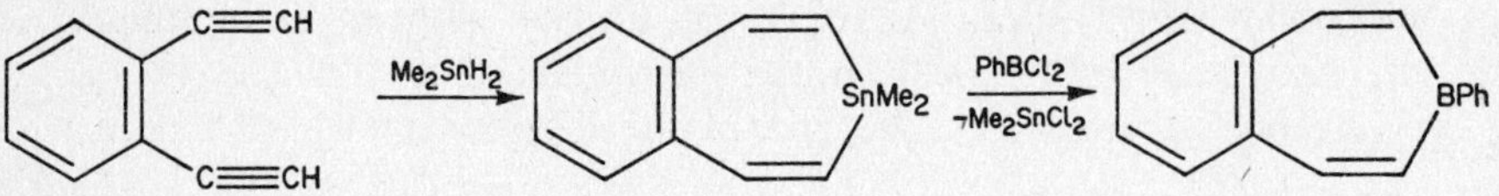

3.8

Co-ordination of strong bases to the boron of 3.8 destroys its aromatic character.

(*e*) *Organoboron hydrides*. Volatile trialkylboranes react with diborane at room temperature to form alkyldiboranes, e.g.

$$2Me_3B + B_2H_6 \rightleftharpoons 2Me_2BHBH_2Me$$

All four terminal hydrogens but not the bridging hydrogens of diborane can be replaced by alkyls, and all five possible methyldiboranes ($MeBH_2BH_3$, Me_2BHBH_3, $MeBH_2BH_2Me$, Me_2BHBH_2Me and $Me_2BHBHMe_2$) are known. The failure to obtain penta- and hexamethyldiborane provided early chemical evidence in favour of the bridge model of diborane.

Alkyldiboranes are labile and tend to disproportionate. They are very reactive to oxygen, water and other protic reagents unless protected by co-ordination with a suitable donor molecule. For example, the hydroboration reagent *t*-butylborane, Bu^tBH_2, is conveniently stored as the air- and moisture-stable trimethylamine adduct, Bu^tBH_2, NMe_3, prepared as follows:

$$B(OMe)_3 \xrightarrow[-78°]{Bu^tMgCl} (Bu^tBO)_3 \xrightarrow[Me_3N]{LiAlH_4} Bu^tBH_2, NMe_3$$

Alkyl derivatives of higher boranes, particularly B_5H_9 and $B_{10}H_{14}$, have been among numerous organoboron compounds investigated as potential high energy fuels. The purpose of alkylation was to obtain fuels with more suitable physical properties (liquid range, vapour pressure, density) than the parent borane without too great a decrease in the calorific value. However, difficulties in exploiting the full oxidation energy of boron superfuels due to incomplete combustion reduced their advantage over conventional fuels to some 40% higher heat content per mass unit, at a cost some 20–50 times greater.

Methods for attaching alkyl groups to the higher boranes include exchange reactions,

$$B_4H_{10} + B_2H_5Me \rightleftharpoons MeB_4H_9 + B_2H_6$$

reactions with olefins in the presence or absence of aluminium chloride,

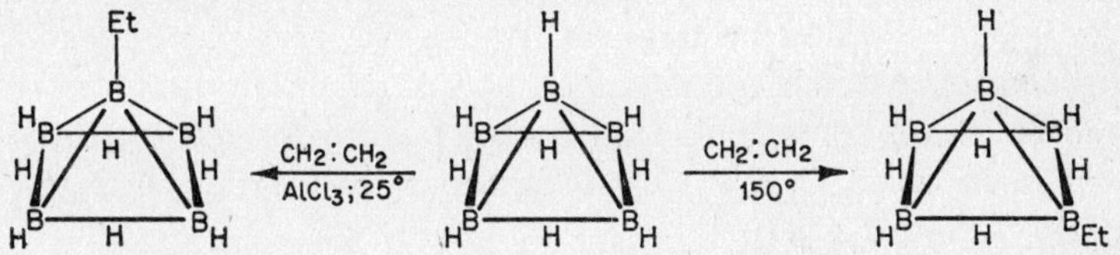

Friedel-Crafts substitution using alkyl halides, or nucleophilic substitution by lithium alkyls. The position of substitution reflects the charge distribution within the borane; electrophilic substitution for example invariably occurs at the apical (i.e. most negative) positions, nucleophilic substitution at the most positive borons:

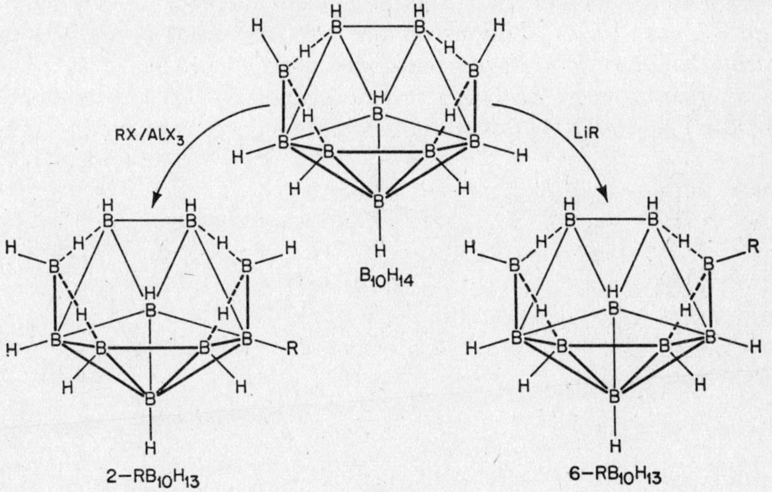

(*f*) *Carboranes*. The organoboron compounds already described have alkyl or aryl groups attached to boron by normal two-centre electron-pair bonds, and it has been pointed out that the absence of electron-deficient bridging by the organic groups of trialkyl- or triaryl-boranes may be a consequence of the unusually short C—C and B—B distances which would be a necessary feature of a BC_2B bridge. That electron-deficient bonding between boron and carbon can occur is evident from the existence of carboranes, cage compounds in which both boron and carbon atoms are incorporated in the electron-deficient framework. Although they were first prepared only a few years ago, it has already become apparent that these compounds have an extensive and fascinating chemistry. Their structures pose interesting valence problems, and afford scope for the study of an effectively three-dimensional electron-delocalized system; the thermal stability and resistance to oxidation or hydrolysis of the $B_{10}C_2$ cage allows a large range of derivatives to be prepared, and has prompted research aimed at incorporating them in useful polymers; moreover, novel transition metal complexes are formed by carborane anions such as $B_9C_2H_{11}^{2-}$ and others of the series $B_nC_2H_{n+2}^{2-}$, $B_{10}CH_{11}^{3-}$ and $B_8C_2H_{10}^{4-}$.

Most of the carboranes known belong to a series of general formula $B_nC_2H_{n+2}$; examples with values of n from 3 to 10 inclusive are known. Several monocarba-derivatives have also been prepared, generally as anions $B_nCH_{n+1}^{-}$; both series are isoelectronic with borane anions $B_mH_m^{2-}$, and have closed cage structures with no bridging hydrogens.

The most readily accessible carboranes are derivatives of $B_{10}H_{10}C_2H_2$, which have structures based on an icosahedral framework in which the two carbon atoms occupy adjacent (*ortho'*) (3.9a), alternate (*'meta'*) (3.9b) or trans (*'para'*) (3.9c) positions on the cage, and each boron and each carbon has one terminal hydrogen atom (not shown in 3.9a, 3.9b and 3.9c). The *ortho* isomer, 1,2-dicarbaclosododecaborane–(12) ('carborane' or 'barene') $B_{10}H_{10}C_2H_2$ ('closo' is used to designate cage boranes which

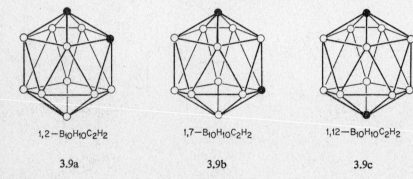

$1,2-B_{10}H_{10}C_2H_2$	$1,7-B_{10}H_{10}C_2H_2$	$1,12-B_{10}H_{10}C_2H_2$
3.9a	3.9b	3.9c

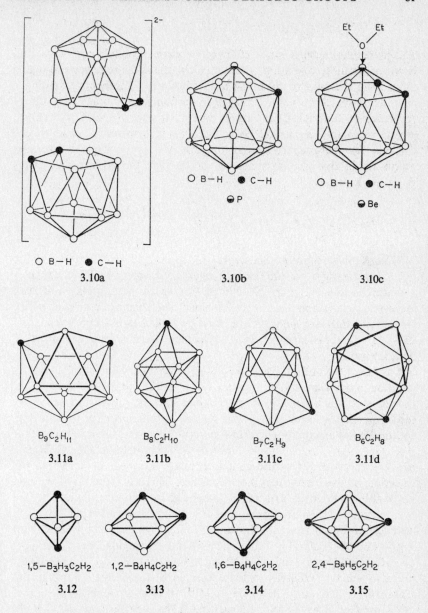

○ B—H   ● C—H
3.10a

○ B—H   ● C—H
⊕ P
3.10b

○ B—H   ● C—H
⊖ Be
3.10c

$B_9C_2H_{11}$
3.11a

$B_8C_2H_{10}$
3.11b

$B_7C_2H_9$
3.11c

$B_6C_2H_8$
3.11d

$1,5-B_3H_3C_2H_2$
3.12

$1,2-B_4H_4C_2H_2$
3.13

$1,6-B_4H_4C_2H_2$
3.14

$2,4-B_5H_5C_2H_2$
3.15

have no bridging hydrogens) and its C-substituted derivatives are readily prepared by the reaction between decaborane–14 derivatives $B_{10}H_{12}L_2$ (where L is a donor molecule such as acetonitrile or diethyl sulphide) and acetylenes:

$$B_{10}H_{14}+HC\!:\!CH \xrightarrow[\text{C}_6\text{H}_6]{\text{2Et}_2\text{S; 90}°} B_{10}H_{10}C_2H_2+2H_2$$

The carborane nucleus is not attacked by water, alcohols and acids, and is very resistant to oxidation, although alcoholic alkali removes one boron atom from the cage to form the anion $B_9C_2H_{12}^{-}$, thereby providing a route to lower carboranes. Chlorine in carbon tetrachloride successively replaces the B-attached hydrogens and then one of the C-hydrogens, giving ultimately $B_{10}Cl_{10}C_2HCl$. Derivatives substituted only at the carbons can be prepared either by use of appropriately substituted acetylenes in the original synthesis, or through the lithium derivatives $B_{10}H_{10}C_2Li_2$:

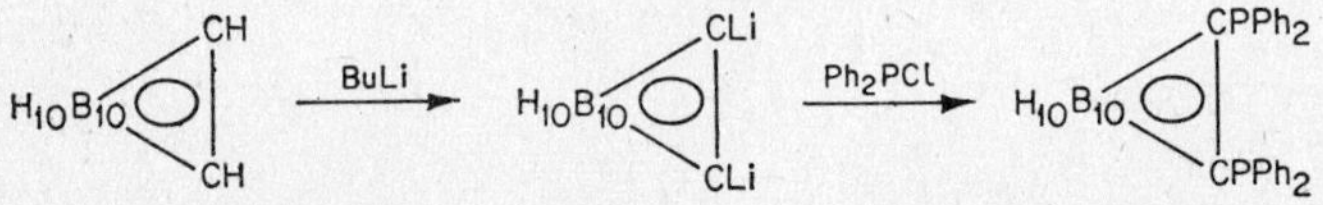

$$H_{10}B_{10}\underset{\text{CH}}{\overset{\text{CH}}{\bigcirc}} \xrightarrow{\text{BuLi}} H_{10}B_{10}\underset{\text{CLi}}{\overset{\text{CLi}}{\bigcirc}} \xrightarrow{\text{Ph}_2\text{PCl}} H_{10}B_{10}\underset{\text{CPPh}_2}{\overset{\text{CPPh}_2}{\bigcirc}}$$

When ortho-carborane, $1,2\text{-}B_{10}H_{10}C_2H_2$, is held at 465–500° for one day, it rearranges to the meta isomer neocarborane, 1,7-dicarbaclosododecaborane–(12), mp 264–266° (3.9b) which is less polar and even more thermally stable than the 1,2 isomer. The chemical reactions of the two compounds are similar. The para or 1,12 dicarba- compound, mp 259–261°, (3.9c) is obtained in 6% yield by the slow rearrangement of neocarborane at 615°.

Treatment of $1,2\text{-}B_{10}H_{10}C_2H_2$ with alcoholic alkali, followed by addition of a tetramethylammonium salt, gives the salt $Me_4N^{+}B_9C_2H_{12}^{-}$, the anion of which is readily converted into $B_9C_2H_{11}^{2-}$ (the 'dicarbollide' ion, from the Spanish olla, = jar) by the abstraction of a proton with sodium metal or sodium hydride in tetrahydrofuran. The structure of $B_9C_2H_{11}^{2-}$ (hydrogen atoms omitted) is illustrated in Figure 20 (a), which shows the sp^3 hybrid atomic orbitals assumed to project in the direction of the empty icosahedral corner. The pentagonal face exposed is capable of bonding to a transition metal in a similar manner to the cyclopentadienide anion. For example, the complex anion $Fe(B_9C_2H_{11})_2^{2-}$ with the sandwich structure shown in Figure 20(b), can be prepared from $FeCl_2$ and $Na_2B_9C_2H_{11}$. Mixed sandwich compounds like $\pi\text{-}C_5H_5Fe\text{-}(\pi\text{-}B_9C_2H_{11})$ (from NaC_5H_5, $Na_2B_9C_2H_{11}$ and $FeCl_2$) Figure 20(c), are also known. Whereas in these complexes the metal completes the icosahedron, derivatives $[(B_9C_2H_{11})_2M]^{2-}$ of d^8 or d^9 metal ions such as Ni(II) or Cu(II) have slipped structures 3.10a in which the metal is no longer located over the centre of the dicarbollide pentagonal face, but is nearer to the three boron atoms in the face, so that the immediate co-ordination sphere of the metal resembles that in bis(π-allyl)nickel (p. 191).

The $B_{10}C_2$ icosahedron can be regenerated from $Na_2B_9C_2H_{11}$ by

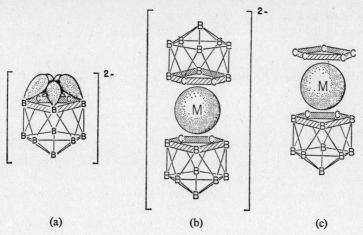

(a)   (b)   (c)

Figure 20.   (a)-(c).   Structures of $B_9C_2H_{11}{}^{2-}$, $Fe(\pi\text{-}B_9C_2H_{11})_2{}^{2-}$, and $\pi\text{-}C_5H_5$, $Fe(\pi\text{-}B_9C_2H_{11})$ with H-atoms omitted and showing the sp^3 hybrid atomic orbitals assumed to be present in $B_9C_2H_{11}{}^{2-}$. (After M. F. Hawthorne and T. D. Andrews, *Chem. Comm.*, 1965, 443).

reaction with $PhBCl_2$, which gives $PhB_{10}C_2H_{11}$. The combination of alkaline degradation and reaction with $PhBCl_2$ thus effects substitution at a boron of $B_{10}C_2H_{12}$. Alkaline degradation of $1,7\text{-}B_{10}C_2H_{12}$ gives anions $B_9C_2H_{12}{}^-$ in which the carbon atoms are no longer adjacent in the icosahedral fragment.

   Although the chemistry of monocarbaclosoboranes is as yet less developed than that of the dicarba- derivatives, they can evidently form similar transition metal complexes. The anion $B_{10}H_{10}CH^{3-}$ for example, being isoelectronic and isostructural with $B_9H_9C_2H_2{}^{2-}$, forms a range of complexes such as $(B_{10}H_{10}CH)_2Ni^{2-}$ or $(B_{10}H_{10}CH)_2Co^{2-}$. Like $B_9H_9C_2H_2{}^{2-}$, $B_{10}H_{10}CH^{3-}$ as a ligand tends to stabilize high oxidation states of transition elements. The following reaction sequence illustrates a route to such derivatives from decaborane:

$$B_{10}H_{14} \xrightarrow[\text{(iii) Me}_2\text{SO}_4]{\text{(i) CN}^- \text{ (ii) H}^+} B_{10}H_{12}CNMe_3 \xrightarrow[\text{THF}]{\text{Na}} B_{10}H_{12}CH^- \xrightarrow[[\text{O}]]{\text{CoCl}_2} (B_{10}H_{10}CH)_2Co^{2-}$$

The anion $B_{10}H_{10}CH^{3-}$ also provides a means of generating icosahedra incorporating other metalloids:

$$Na_3B_{10}H_{10}CH + PCl_3 \longrightarrow 3NaCl + B_{10}H_{10}CHP \qquad (3.10b)$$

A further method of incorporating another element in the icosahedron is illustrated by the acid reaction of $B_9C_2H_{13}$ on Me_2Be:

$$B_9C_2H_{13} + Me_2Be \xrightarrow{\text{Et}_2\text{O}} Et_2O, BeB_9C_2H_{11} \qquad (3.10c)$$

The compound $B_9C_2H_{13}$, which has bridging hydrogens and is accordingly much less thermally stable than the B_{10} carboranes, is a key intermediate in the preparation of lower carboranes $B_nC_2H_{n+2}$, where $n = 6$, 7, 8 or 9

$$B_{10}C_2H_{12} \xrightarrow[\text{(ii) H}^+]{\text{(i) ROH/OH}^-} B_9C_2H_{13} \xrightarrow{130°} B_9C_2H_{11} \xrightarrow[\text{aq. HOAc}]{\text{Cr}_2\text{O}_7{}^{2-}} B_7C_2H_{13} \xrightarrow{200°} B_nC_2H_{n+2}$$

$$(n = 6, 7 \text{ or } 8)$$

The cage structures of these lower carboranes and their derivatives are exemplified by 3.11a-d.

Reactions between acetylenes and lower boron hydrides afford routes to other carboranes. Carboranes with three, four or five boron atoms for example can be prepared *via* dicarbahexaborane–(8), $B_4C_2H_8$, which results from the reaction between B_5H_9 and acetylene at elevated temperatures:

$$1,5-B_3H_3C_2R_2 + 1,2-B_4H_4C_2R_2$$

$$B_5H_9 \xrightarrow{RC\vdots CR} B_4C_2H_6R_2 \xrightarrow[\text{or electric discharge}]{\text{pyrolysis, u.v.}}$$

$$+1,6-B_4H_4C_2R_2 + 2,4-B_5H_5C_2R_2$$

These lower carboranes, which have the bipyramidal cage frameworks illustrated in 3.12, 3.13, 3.14 and 3.15, are less reactive than compounds with $B\cdots H\cdots B$ bridges. The carborane $1,5-B_3H_3C_2H_2$, a colourless gas, bp $-4°$, for example, is thermally stable at $25°$ at which temperature it does not react with acetone, trimethylamine, air or water, in striking contrast to the lower boranes.

Various anionic derivatives of lower carboranes, containing from six to nine cage atoms, have been prepared and found to form stable transition metal complexes, e.g.

$$MeC_3B_3H_6 + \tfrac{1}{2}Mn_2(CO)_{10} \xrightarrow{175-200°} (MeC_3B_3H_5)Mn(CO)_3 \qquad (3.15a)$$

$$B_7C_2H_{13} \xrightarrow{2NaH/Et_2O} Na_2B_7C_2H_{11} \xrightarrow{CoCl_2/Et_2O} NaCo(B_7C_2H_9)_2 \qquad (3.15b)$$

$$Co(B_9C_2H_{11})_2{}^- \xrightarrow{30\%\,\text{aq.NaOH/CoCl}_2/100°} [(B_9C_2H_{11})Co(B_8C_2H_{10})Co(B_9C_2H_{11})]^{2-} \qquad 3.15c$$

A common feature of all the carboranes is the high co-ordination numbers (generally five or six) of the borons and carbons of the cages. The bonding cannot therefore be described satisfactorily in terms of localized electron pair two-centre bonds, and even bonding schemes invoking three-centre bonds such as are adequate for a description of bridged alkyls of Group II or Group III elements, or indeed of many boron hydrides, are unsatisfactory when applied to carboranes. Molecular orbital treatments taking into account the symmetry of the complete molecule are much more successful. For the icosahedral cage of the

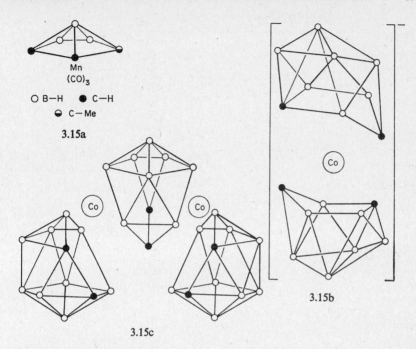

Mn
(CO)$_3$

○ B—H   ● C—H
◐ C—Me

3.15a

3.15b

3.15c

$C_2B_{10}H_{12}$ carboranes, for example, interactions between the atomic orbitals of the constituent borons and carbons can be shown to lead to a set of 13 bonding molecular orbitals within the cage. As there are 26 electrons available apart from those involved in terminal B—H or C—H bonds, these are just sufficient to fill all the bonding orbitals.

(g) *Organoboron-nitrogen compounds.* Boron-nitrogen compounds have a particular interest arising from their relationship to organic compounds. When a C—C link in an organic compound is replaced by a B—N link, the resulting compound is iso-electronic with, but more polar than, the original (e.g. compare CH_3CH_3 with BH_3NH_3). Although structurally there is generally a close correspondence between boron-nitrogen compounds and their organic analogues, the greater polarity of the former is normally reflected in their greater reactivity to electrophiles or nucleophiles. Several classes of boron-nitrogen compound are listed in Table VI, together with their organic counterparts.

Borane adducts R_3N,BR_3 and boronium salts $[(R_3N)_2BR_2]^+X^-$, formal analogues of alkanes, are described together with other co-ordination compounds on pp 99–104.

Aminoboranes $(R_2BNR_2)_n$ if monomeric contain three co-ordinate boron and nitrogen and are formally analogous to alkenes; if associated ($n = 2$ or 3) they contain four co-ordinate boron and nitrogen and struc-

G

Table VI. *Boron-nitrogen analogues of organic systems.*

Organic compound		Boron-nitrogen analogue	
Alkane	$R_3C\!-\!CR_3$	$R_3N \longrightarrow BR_3$	borane adduct (borazane)
	$(R_3C)_2CR_2$	$(R_3N)_2BR_2^+X^-$	boronium salt
alkene	$R_2C \!=\! CR_2$	$R_2N \rightleftharpoons BR_2$	monomeric aminoborane (monomeric borazene)
cycloalkane	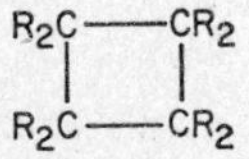 $\begin{array}{c} R_2C\!-\!CR_2 \\ \mid\quad\ \ \mid \\ R_2C\!-\!CR_2 \end{array}$	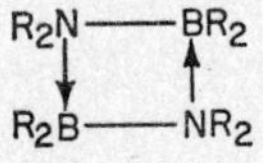 $\begin{array}{c} R_2N\!-\!BR_2 \\ \uparrow\quad\ \ \downarrow \\ R_2B\!-\!NR_2 \end{array}$	dimeric aminoborane (cycloborazane)
	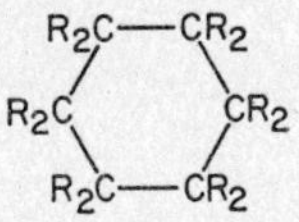 six-membered carbon ring R_2C/CR_2	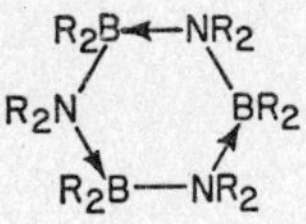 six-membered B–N ring $R_2B{\leftarrow}NR_2$ / R_2N / BR_2 / $R_2B{-}NR_2$	trimeric aminoborane (cycloborazane)
alkyne	$RC \equiv CR$	$RN \overset{\rightleftharpoons}{} BR$	monomeric borazyne
allene	$R_2C \!=\! C \!=\! CR_2$	$R_2C \!=\! N \rightleftharpoons BR_2$	monomeric ketiminoborane
cyclobutadiene	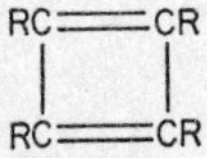 $\begin{array}{c} RC\!=\!CR \\ \mid\quad\ \ \mid \\ RC\!=\!CR \end{array}$	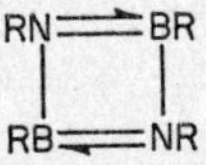 $\begin{array}{c} RN\rightleftharpoons BR \\ \mid\qquad\ \mid \\ RB\rightleftharpoons NR \end{array}$	boretane (dimeric borazyne)
benzene	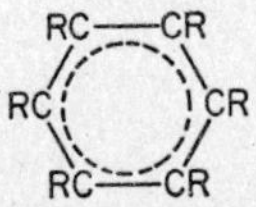 aromatic six-membered carbon ring (RC)$_6$	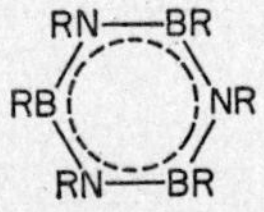 aromatic six-membered ring: $RN\!-\!BR$ / RB / NR / $RN\!-\!BR$	borazine, borazole (trimeric borazyne)
		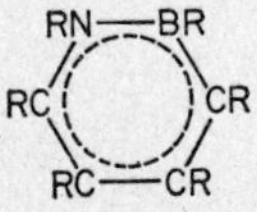 $RN\!-\!BR$ / RC / CR / $RC\!-\!CR$	borazarene
cyclooctatetraene	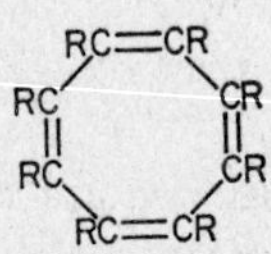 eight-membered carbon ring $RC\!=\!CR$ (×4)	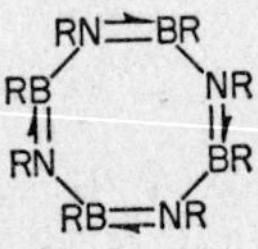 eight-membered ring $RN\rightleftharpoons BR$, RB, NR, RN, BR, $RB\!=\!NR$	borazocine (tetrameric borazyne)

turally resemble cycloalkanes. They are commonly prepared from boron halides:

$$R_2BCl \xrightarrow[\text{or } R_2NH + Et_3N]{\text{LiNR}_2 \text{ or Me}_3\text{SiNR}_2} R_2BNR_2 \xleftarrow[\text{(ii) } 2RMgX]{\text{(i) } 2R_2NH} BCl_3$$

Their state of association (and therefore their reactivity) appears to be influenced by *steric, mechanistic* and *electronic* effects. Bulky substituents may prevent association. For example, the compound Cl_2BNPr_2 is a monomeric hydrolytically sensitive liquid, whereas Cl_2BNMe_2 can exist as a very reactive liquid, violently hydrolysed by cold water, which slowly changes into an unreactive solid dimer. Association to form trimers, which have chair-shaped rings like that of cyclohexane, involves even greater crowding of the substituents, and is found in only a few examples including cyclotriborazane itself $(H_2BNH_2)_3$. Where trimers are sterically possible, they are apparently formed in preference to dimers in reactions involving oligomeric or polymeric (rather than monomeric) intermediates, e.g.

$$H_3B,NH_2Me \longrightarrow [MeH_2N \cdot BH_2 \cdot NHMe \cdot BH_2 \cdot NH_2Me]^+BH_4^- \longrightarrow (H_2BNHMe)_3$$

Electronic effects of substituents may affect the state of association of aminoboranes by reducing the acceptor properties of the boron or donor properties of nitrogen. Double bonding is favoured by electropositive groups on boron, as shown by the dissociation of $(Me_2BNH_2)_2$ in the vapour phase, and by the monomeric nature of Me_2BNMe_2.

In monomeric aminoboranes, rotation about the double $B \rightleftharpoons N$ bond is hindered. This can be demonstrated readily by the ^1H n.m.r. spectra of compounds Me_2BNRR' or $XYBNMe_2$, in which the methyl groups differ, with energy barriers to rotation of ca. 14–18 kcal/mole (cf. ca. 60 kcal/mole for rotation about the $C=C$ bond of comparable alkenes). Moreover, derivatives $RR'BNRR'$ can exist in *cis* and *trans* forms, although interaction between the substituents may make the *trans* isomer more stable. *Cis- trans*-isomerism may also occur in dimeric aminoboranes $(X_2BNHR)_2$ (3.16) or aldiminoboranes, e.g. $(MeCH:N \cdot BMe_2)_2$ (3.17 and 3.18).

3.16   3.17   3.18

Monomeric ketiminoboranes $Ph_2C=N \rightleftharpoons BR_2$ (from $Ph_2C:NLi + R_2BCl$) are formal analogues of allenes $Ph_2C:C:CR_2$, and are characterized by bands near 1800 cm^{-1} in their infrared spectra, indicating a linear $C=N \rightleftharpoons B$ skeleton.

Among other boron-nitrogen analogues of unsaturated organic compounds, the borazines or borazoles $(RBNR')_3$ have attracted much attention since the parent compound (3.19) was first prepared in 1926 in the course of A. Stock's classic researches on boron hydrides. Borazines and benzene have very similar physical properties and structures, and the $(BN)_3$ ring is a regular hexagon. Its aromatic character is however, slight, the

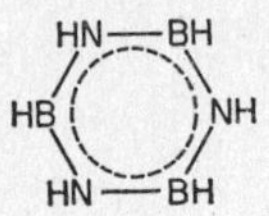

3.19 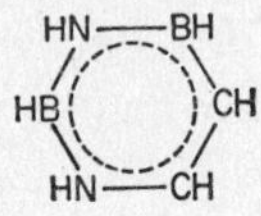   3.20 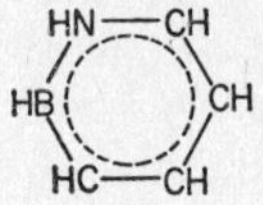   3.21

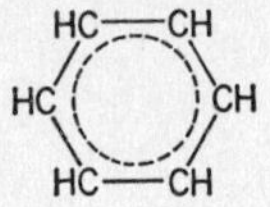

electron density alternating round the ring because of the different electronegativities of the boron and nitrogen atoms.

Borazines are generally prepared by condensations between amines and boron hydrides or halides at up to 200°, with elimination of hydrogen or hydrogen halide:

$$3BX_3 + 3R_3N \longrightarrow X_3B_3N_3R_3 + 6RX$$

(X = H, Cl, Br, Alk; R = H, Alk). A convenient laboratory preparation of $Cl_3B_3N_3H_3$ for example uses the action of ammonium chloride on boron trichloride (preferably as $MeCN,BCl_3$) in chlorobenzene; reduction then affords borazine itself.

$$3MeCN,BCl_3 + 3NH_4Cl \xrightarrow[\text{PhCl}]{\text{boiling}} Cl_3B_3N_3H_3 \xrightarrow[\text{diglyme}]{\text{NaBH}_4} H_3B_3N_3H_3$$

B-Alkyl or -aryl borazines are obtained by the action of Grignard reagents on the chloride.

Borazines are much more reactive than analogous benzene compounds. A marked difference from benzene which reflects the polarity of the ring bonds is that borazine adds water, methanol, alkyl iodides and hydrogen halides, the negative parts of which become attached to the borons:

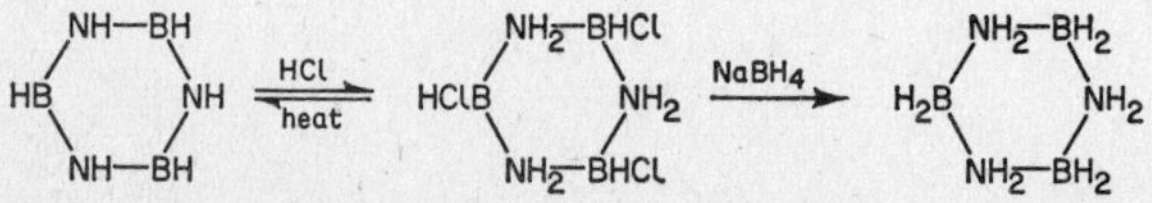

Bulky substituents hinder such reactions and, consequently, hydrolysis.

Borazines form π-complexes with transition metals (p 210).

In the sense that borazine $(HBNH)_3$ may be regarded as derived from benzene by the replacement of all six carbons by three BN units, it represents the limiting member of a series in which the carbon atoms of benzene are replaced in pairs by BN units. Such rings as $C_2B_2N_2$ (3.20) and C_4BN (3.21) should also have some aromatic character.

Many derivatives of 3.21 have been prepared recently, largely by M. J. S. Dewar and his co-workers, and shown to have typical aromatic properties (characteristic electronic spectra, susceptibility to substitution reactions, resistance of the B—N bond to hydrolytic cleavage). Accordingly, they are called borazarenes. Many have been prepared by Friedel-Crafts reactions (p 75) or other condensations of suitable aminoboranes:

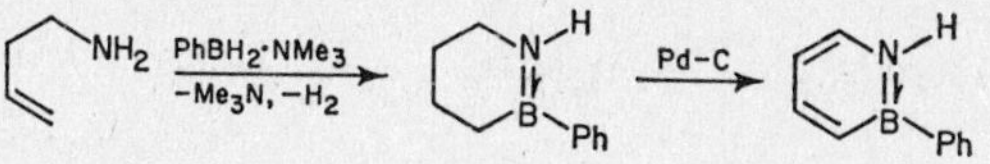

Most are multi-ring systems whose chemistry is beyond the scope of this book.

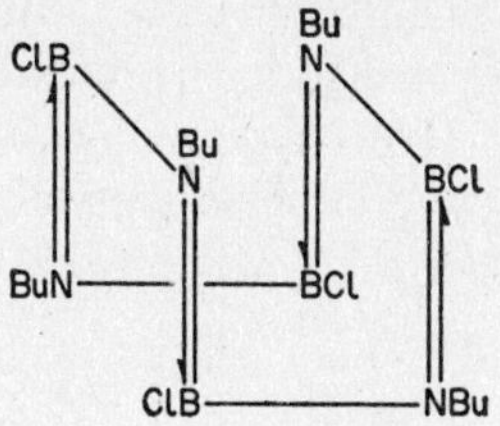

3.22 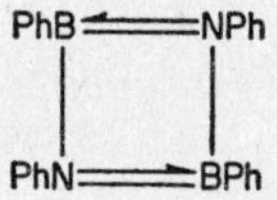   3.23

In borazines and borazarenes, units RBNR′ ('borazyne' units) are incorporated in six-membered rings. Oligomeric borazynes $(RBNR')_n$ other than trimers are obtained if the substituents R and R′ are too bulky to be accommodated about a planar six-membered ring. Thus, Bu^tNH_2 and BCl_3 form the tetrameric borazyne (borazocine) $(ClBNBu^t)_4$, which has a boat-shaped ring (3.22) like that of cyclo-octatetraene. Dimeric borazynes (boretanes, 3.23, cf. cyclobutadienes) result from rearrangement reactions of certain azidoboranes:

$$LiN_3 + Ph_2BCl \longrightarrow Ph_2BN_3 \xrightarrow{-N_2} [PhBNPh] \longrightarrow (PhBNPh)_2$$

These rearrangements apparently involve monomeric borazynes as intermediates, and indeed air- and moisture-sensitive monomeric borazynes $RB{\equiv}NR'$ result from the thermal decomposition of certain aryl amine adducts of $C_6F_5BCl_2$:

$$C_6F_5BCl_2 + ArNH_2 \xrightarrow{60°;\ C_6H_6} C_6F_5B{\equiv}NAr + 2HCl$$

Formal analogues of alkynes, these compounds have a high BN bond order as shown by $\nu(BN)$ (1700 cm^{-1}) in their vibrational spectra (cf. borazines, $\nu(BN)$ 1375–1495 cm^{-1}, and amine-boranes R_3B,NR_3, $\nu(BN)$ 650–800 cm^{-1}).

(h) *Organoboron-oxygen compounds.* Boron forms two series of organic acids, borinic acids R_2BOH and boronic acids $RB(OH)_2$. Borinic acids are easily obtained by the hydrolysis of halides R_2BX or (aryls particularly) of aminoboranes R_2BNPh_2

$$BCl_3 \xrightarrow{Ph_2NH} Cl_2BNPh_2 \xrightarrow{RMgX} R_2BNPh_2 \xrightarrow[H_2NC_2H_4OH]{H_2O} R_2B\begin{array}{c} O \\ \diagup \diagdown CH_2 \\ \diagdown \diagup \\ H_2N \quad CH_2 \end{array} \xrightarrow{HCl} R_2BOH$$

3.24

They are conveniently isolated and stored as ethanolamine esters, which owe their air-stability to their cyclic boroxazolidine structure (3.24) in which the four co-ordinate boron is protected from attack by nucleophiles such as water or molecular oxygen. 8-Hydroxyquinoline, β-ketoesters and β-keto enols form similar derivatives, and their acetylacetonates (3.25) result from the smooth cleavage of one alkyl group from trialkylboranes R_3B:

$$R_3B + MeCOCH_2COMe \longrightarrow R_2B\begin{array}{c} O—CMe \\ \diagup \diagdown \\ + \quad CH \\ \diagdown \diagup \\ O—CMe \end{array} \xrightarrow[(MeCO)_2CH_2]{hydrolysis} R_2BOH$$

3.25

Boronic acids $RB(OH)_2$ and anhydrides $(RBO)_3$ are more easily prepared and less sensitive to oxidation than borinic acids or anhydrides. The direct Grignard reaction can be used:

$$(MeO)_3B \xrightarrow{RMgX} RB(OMe)_2 \xrightarrow{H_2O} RB(OH)_2 \rightleftharpoons 1/3\,(RBO)_3 + H_2O$$

Boronic acids are easily dehydrated to boroxines, the Dean and Stark method (distil off the water with benzene) being quite satisfactory except for the volatile methyl and ethyl compounds.

Phenylboronic acid $PhB(OH)_2$ is cleaved by halogens, acids and mercury halides under mild conditions in aqueous solution:

$$PhB(OH)_2 + HgCl_2 + H_2O \longrightarrow PhHgCl + HCl + H_3BO_3$$

Studies on the kinetics of such deboronations and on other substitution reactions of this type (e.g. desilylation, degermylation and destannylation) have led to a better understanding of the mechanism of electrophilic aromatic substitutions, and have extended the range of the 'selectivity principle' in such reactions. Aprotic diazotization of ortho-aminophenyl-boronic acid generates benzyne:

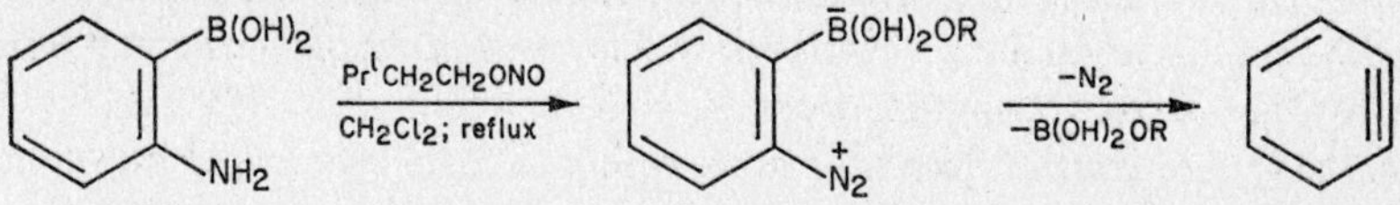

Like borinic acids, boronic acids are weak acids and are readily characterized as cyclic derivatives such as diethanolamine esters (3.26) pyridine adducts of their anhydrides (3.27) or as benzodiazaboroles (3.28).

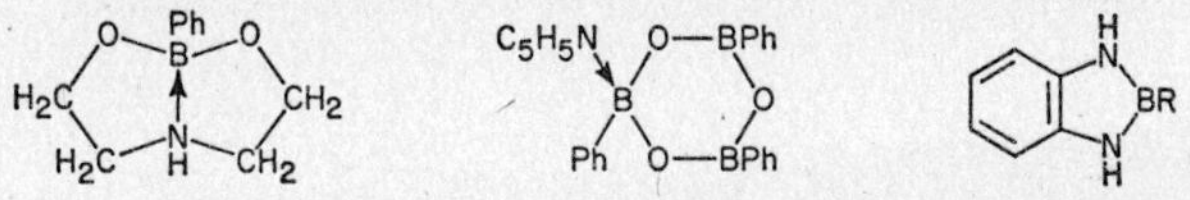

Boroxines (3.29), the cyclic anhydrides of boronic acids, are of special interest in view of their possibly aromatic character. The parent compound (3.29)

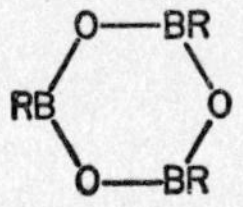

(R = H) is unstable with respect to boric oxide and diborane at ordinary temperatures, but can be obtained by treating diborane with oxygen and caution at low pressure, the reaction being initiated by a spark from a Tesla coil. Alkylboroxines result from the dehydration of boronic acids, but are more conveniently prepared from trialkylboranes and B_2O_3 or $(MeOBO)_3$. Calculations indicate that their aromatic character is slight—their π-electronic structure is more localized for example than that of borazines. Little aromatic character is observable in boroxarenes, boron-oxygen analogues of the borazarenes already described (p. 87) which are prepared by similar reactions:

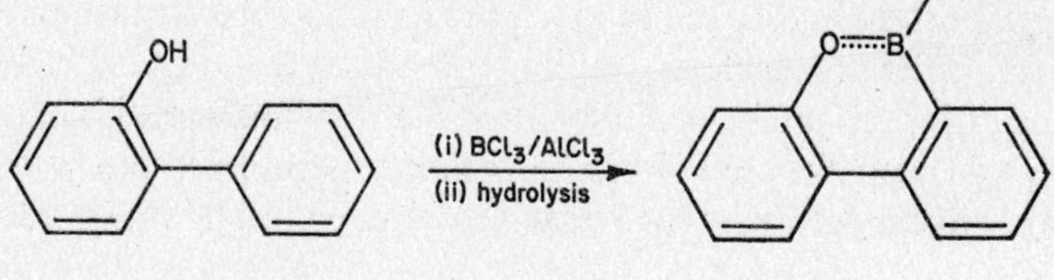

Their aromatic character for example is insufficient to prevent some derivatives being quite rapidly oxidized by air.

(*i*) *Compounds related to* B_2X_4. Although organic substituents have a stabilizing influence on catenated compounds of Group IV elements (p 124), a different situation obtains with boron. The thermal stability of compounds $X_2B \cdot BX_2$ *decreases* in the order $X = R_2N > RO \sim RS > F > Cl > R > H$, that is as the group X decreases in its ability to form π-bonds to boron and so relieve the co-ordinative unsaturation. Organoboron compounds of this type are stable only if amino groups are also present, as in derivatives $R(Me_2N)BB(NMe_2)R$ (from $BuBCl(NMe_2) + Na/K$) which are stable to $100°$. The boron-boron link in such compounds is strong enough to survive a transamination reaction with *o*-phenylenediamine:

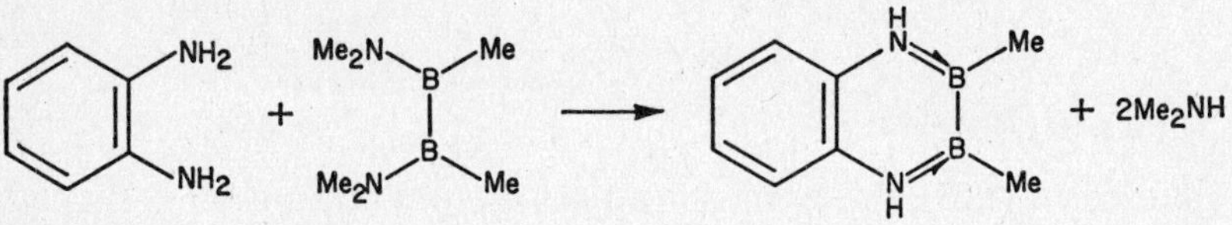

Catechol in contrast causes disproportionation:

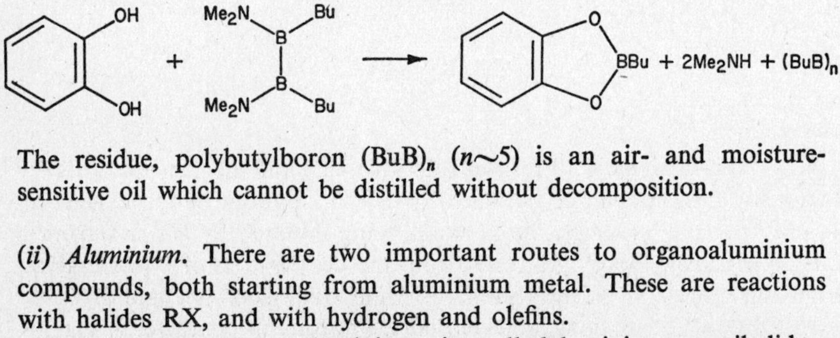

The residue, polybutylboron $(BuB)_n$ $(n \sim 5)$ is an air- and moisture-sensitive oil which cannot be distilled without decomposition.

(*ii*) *Aluminium*. There are two important routes to organoaluminium compounds, both starting from aluminium metal. These are reactions with halides RX, and with hydrogen and olefins.

With suitable halides, aluminium gives alkylaluminium sesquihalides:

$$2Al + 3RX \longrightarrow R_3Al_2X_3$$

This reaction is very limited in scope, but gives some useful products. It works well with methyl and ethyl halides, the chlorides being much the most important as they are so much cheaper than bromides and iodides. Aluminium generally needs activating, and one of the most effective ways is by means of some preformed alkylaluminium compound, though iodine or $AlCl_3$ can also be used. Phenylaluminium sesquichloride is formed, only with difficulty, when PhCl reacts with aluminium which has been activated by grinding it with $AlCl_3$. Propyl and higher alkyl halides give, mainly, $AlCl_3$ and hydrocarbons on reaction with aluminium.

The sesquihalides disproportionate

$$2\,R_3Al_2X_3 \rightleftharpoons R_2Al_2X_4 + R_4Al_2X_2$$

and the process continues in the case of $Me_3Al_2I_3$ as far as Me_3Al, which can thus be made in moderate yield from methyl iodide and aluminium. Such a preparation would be quite uneconomical industrially, but is sometimes undertaken in the laboratory when trimethylaluminium cannot readily be obtained from industrial sources.

Methylaluminium sesquichloride reacts very cleanly with salt,

$$2Me_3Al_2Cl_3 + 2NaCl \longrightarrow 2NaMeAlCl_3 + Me_4Al_2Cl_2$$

and the $Me_4Al_2Cl_2$ can be poured from the complex halide and then distilled. It or $Me_3Al_2Cl_3$ can then be reduced with sodium (as in one of the industrial routes to Me_3Al):

$$3Me_4Al_2Cl_2 + 6Na \longrightarrow 6NaCl + 2Al + 4Me_3Al$$

The synthesis of aluminium alkyls from metal, hydrogen, and olefin has been mentioned briefly in Chapter 2, and is used on a far bigger scale than the processes involving methyl or ethyl chloride, mainly for the manufacture of triethyl-. tri-isobutyl-, and to a lesser extent of tri-n-propyl-aluminium.

The key to the success of this method, which is much more suitable to industrial than to laboratory use, is the production of metal with an active surface. One way of achieving this is ball-milling the powder in contact with the metal alkyl.

The two main discoveries, both due to K. Ziegler to whom a Nobel Prize has been awarded, were first that Al—H bonds add to olefins, particularly in the absence of ether and other bases, as well as to carbonyl and similar groups (the latter reactions had been discovered a few years earlier), and second, that although aluminium metal does not form AlH_3 by direct reaction with hydrogen, it does take up hydrogen *in the presence of aluminium alkyl*:

$$Al + 3/2H_2 + 2Et_3Al \longrightarrow 3Et_2AlH$$

Application of the first of the above reactions then gives Et_3Al:

$$3Et_2AlH + 3C_2H_4 \longrightarrow 3Et_3Al$$

Thus the overall process amounts to:

$$Al + 3/2H_2 + 3C_2H_4 \longrightarrow Et_3Al$$

Pure aluminium does not undergo this reaction. A trace of transition metal impurity, generally titanium, is a necessary catalyst. Tri-isobutyl-aluminium is normally made in a single-stage process in which the metal reacts with liquid isobutene and hydrogen at 200 atmospheres (80–110°), again starting with some pre-formed Bu^i_3Al. This is one of the most useful of the aluminium alkyls as many others can be made from it by displacement reactions. Though it distils unchanged if the temperature is kept below 50° (pressure ca. 0·05 mm), it loses isobutene smoothly when heated at 140° and about 20 mm.

$$Bu^i_3Al \longrightarrow Bu^i_2AlH + Me_2C{:}CH_2$$

The reaction is, of course, reversible and Bu^i_3Al is formed again if Bu^i_2AlH is heated to 60–70° with isobutene under mild pressure.

If Bu^i_3Al is heated with propene, then isobutyl are displaced by propyl groups, and ethylene similarly displaces propene:

$$Bu^i_3Al + 3C_3H_6 \longrightarrow Pr^n_3Al + 3Me_2C{:}CH_2$$

$$Pr^n_3Al + 3C_2H_4 \longrightarrow Et_3Al + 3C_3H_6$$

These reactions are believed to involve the fission of an Al—C bond giving an Al—H bond, followed by the addition of the latter to the olefin. The affinities of the three types of olefin for addition to Al—H are

$$C_2H_4 > MeCH{:}CH_2 > Me_2C{:}CH_2$$

Equilibrium constants for the reactions in which propene (or $RCH{:}CH_2$ in general) displace isobutene (or $R_2C{:}CH_2$ in general), by reversible processes involving an R_2AlH intermediate, are about 40, and the equilibrium constants for the displacement of $RCH{:}CH_2$ by C_2H_4 are about the same.

Since Al—H bonds are polarized, all three types of olefin give alkyls containing $\alpha\text{-}CH_2$ groups:

$$
\begin{array}{c}
\overset{\delta+}{R_2 Al} {-} \overset{\delta-}{H} \\
CH_2 {=} CH{\cdot}Me \\
\overset{\delta-}{} \quad \overset{\delta+}{}
\end{array}
\longrightarrow
\begin{array}{c}
R_2 Al \\
| \\
CH_2 {-} CH_2 Me
\end{array}
$$

This is a very important reaction since it allows the preparation of a wide range of aluminium alkyls of the type $(RCH_2CH_2)_3Al$, simply by heating Bu^i_3Al with $RCH{:}CH_2$. The reactions with ethylene and C_3H_6 have already been mentioned, and Et_3Al can be made this way, with circulation of the isobutene. Longer hydrocarbon chains can be introduced even more easily, e.g.

$$3C_8H_{17}CH{:}CH_2 + Bu^i_3Al \longrightarrow (C_{10}H_{21})_3Al + 3Me_2C{:}CH_2$$

and n-decyl groups can then be transferred to other molecules by a variety of Grignard-like alkylation procedures.

If triethyl-aluminium is heated with ethylene at 90–120° and about 100 atmospheres pressure then olefin molecules are slowly inserted into the Al—C bonds in the trialkyl:

$$\text{Al} \begin{array}{l} {\diagup} C_2H_5 \\ {\longleftarrow} C_2H_5 \\ {\diagdown} C_2H_5 \end{array} \xrightarrow{C_2H_4} \text{Al} \begin{array}{l} {\diagup} CH_2{\cdot}CH_2{\cdot}C_2H_5 \\ {\longleftarrow} C_2H_5 \\ {\diagdown} C_2H_5 \end{array} \xrightarrow{C_2H_4} \text{Al} \begin{array}{l} {\diagup} (C_2H_4)_m{\cdot}C_2H_5 \\ {\longleftarrow} (C_2H_4)_n{\cdot}C_2H_5 \\ {\diagdown} (C_2H_4)_o{\cdot}C_2H_5 \end{array}$$

Kinetic studies indicate that monomeric Et_3Al reacts (not Et_6Al_2). Trimethylaluminium does not add olefin in this way. The insertion process is in competition with fission into al$-$H$+$olefin, a process whose rate does not depend appreciably on pressure and which results in the immediate conversion of al—H into al—Et on account of the high affinity of C_2H_4 for al—H (it is convenient to represent 1/3 Al as al). The 'growth' reaction goes faster as the pressure is raised, so high pressures favour larger values of m, n and o in the resulting trialkyl. At about 100 atmospheres the alkyl chains can be grown to about C_{200} before they become detached as long-chain olefins, but this is possible only with very careful exclusion of impurities some of which (particularly nickel) strongly catalyse the displacement reaction. Hydrolysis of the high-molecular-weight aluminium alkyls made this way gives unbranched long-chain hydrocarbons, necessarily containing an *even* number of carbon atoms. However, the production of medium- to high-molecular-weight 'linear' polyethylene by such a route is very slow, and such material is more satisfactorily obtained by the use of the aluminium-alkyl-transition metal catalysts discussed on p 95.

The rate of olefin insertion can be greatly increased if temperatures up to 160° are used, together with short contact times. When carried out in such a way as to produce an average chain length of about C_{14}, this is the basis of a very important industrial process since oxidation of the product by molecular oxygen gives alcohols:

$$\text{al}-C_2H_5 \xrightarrow{C_2H_4} \text{alC}_{\sim14}H_{\sim29} \xrightarrow{O_2} \text{alOC}_{\sim14}H_{\sim29} \xrightarrow{H_2O} \text{alOH}+C_{\sim14}H_{\sim29}OH$$

Unbranched aliphatic alcohols about C_{14} are much in demand for the manufacture of bio-degradable detergents, and plants capable of making about 150 000 tons a year of such products by the aluminium alkyl route have been built in the United States. This application alone puts organo-aluminium compounds on much the same scale of industrial manufacture as silicone polymers.

If the terminal olefins are needed, instead of the alcohols, these can be obtained by carrying out the growth reaction to the desired average chain length, and then adding nickel to catalyse the displacement reaction:

$$\text{al}(C_2H_4)_nC_2H_5 \xrightarrow{Ni} \text{alH}+C_2H_5(C_2H_4)_{n-1}CH{:}CH_2$$

followed immediately by

$$alH + C_2H_4 \longrightarrow alC_2H_5$$

Alternatively, the terminal olefin may be displaced by a very brief (ca. 1 sec) treatment with ethylene at 300°.

Quite different results are obtained when olefins other than ethylene react with aluminium alkyls. If propene and Pr^n_3Al are heated at 140–200° under pressure, then an insertion reaction takes place as with ethylene.

$$Pr_2^nAl\!-\!Pr^n + MeCH\!:\!CH_2 \longrightarrow Pr_2^nAlCH_2\overset{\displaystyle Me}{\underset{\displaystyle Pr^n}{\overset{|}{\underset{|}{C}}}H}$$

In this case only *one* olefin molecule is inserted because the product has a branch at the β-carbon, like $al\!\cdot\!CH_2CHMe_2$, and like Bu^i_3Al, very rapidly loses an olefin of the type $CH_2\!:\!CR_2$,

$$alCH_2\overset{\displaystyle Me}{\underset{\displaystyle Pr^n}{\overset{|}{\underset{|}{C}}}H} \longrightarrow al\!-\!H + CH_2\!:\!\overset{\displaystyle Me}{\underset{\displaystyle Pr^n}{\overset{|}{\underset{|}{C}}}}$$

followed by the addition of propene to alH,

$$alH + CH_2\!:\!CHMe \longrightarrow alPr^n$$

The overall process, therefore,

$$Pr^n_3Al + 2C_3H_6 \longrightarrow Pr^n_3Al + CH_2\!:\!CMePr^n$$

consists of the catalytic dimerization of the propene. Triethyl- or tri-isobutyl-aluminium would give the same result since both would quickly be converted to Pr^n_3Al. All α-olefins can be dimerized in this way:

$$2RCH\!:\!CH_2 \longrightarrow RCH_2CH_2CR\!:\!CH_2$$

The aluminium-alkyl-catalyzed dimerization of propene, followed by thermal elimination of methane, is one of the major industrial processes for making isoprene (itself subsequently polymerized by a transition metal aluminium alkyl catalyst):

$$2MeCH\!:\!CH_2 \xrightarrow{\;R_3Al\;} CH_3CH_2CH_2\underset{\displaystyle CH_3}{\overset{|}{C}}\!:\!CH_2 \xrightarrow{\;cracking\;} CH_2\!:\!CH\!\cdot\!\underset{\displaystyle CH_3}{\overset{|}{C}}\!:\!CH_2 + CH_4$$

The catalytic polymerization of α-olefins by transition metal catalysts is rather a different process. A typical catalyst (for C_2H_4) is obtained from Et_3Al and $TiCl_4$ in heptane; the resulting chocolate-brown suspension absorbs a large amount of ethylene at room temperature and atmospheric pressure with the formation of 'linear' polyethylene, mp 130–135° (contrast the lower-melting, softer, and lower density branched-chain polythene obtained by the high pressure process). The titanium (IV) is reduced and the catalyst is a compound of titanium in a lower valency state. Catalysts can be prepared from the α form of $TiCl_3$ (and various other transition metal compounds), and aluminium (or other electropositive metal) alkyl. Many of these catalysts (developed in part by G. Natta) cause the stereoregular polymerization of propene and other α-olefins. The crystalline isotactic polypropene, in which successive C_3 units have the same stereochemical configuration, melts at about 170°, whereas the disordered (or atactic) polymer is an oil at room temperature. It is much more difficult to establish the mechanism of heterogeneous catalysis than that of homogeneous catalysis (for an example of a tolerably well-understood example of the latter, consider the dimerization of propene by Pr^n_3Al), but it seems likely that the function of the aluminium alkyl (or $R_3Al_2X_3$) is to maintain some alkylated titanium atoms at the surface of, say, a $TiCl_3$ crystal. The olefin co-ordinates to a vacant site, and it can be shown that it would always have to take up the same position relative to the existing Ti—R group. The process can be represented thus (in simplified form):

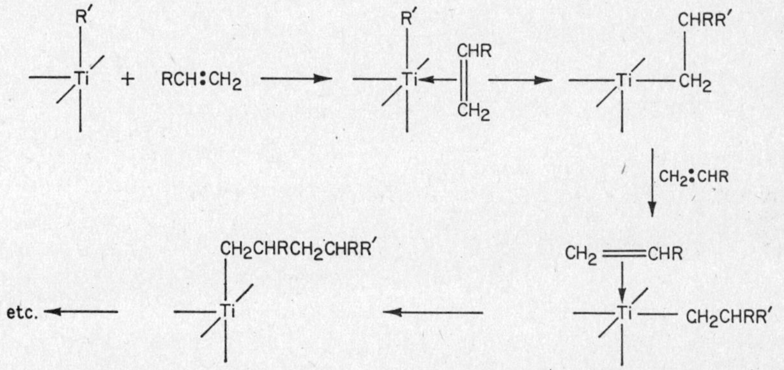

The polymerization process is inhibited by the presence of bases, particularly those of 'B' character which complex well to transition metals in low oxidation states, since the active co-ordination sites are then occupied. An alternative mechanism involves the formation of ion-pairs $[TiCl_2^+]$ $[R_3AlCl^-]$, in which the olefin also co-ordinates to the titanium in much the same way as outlined above, but after the formation of the Ti—C bond the organic chain is transferred to Al. The essential feature in both cases is the co-ordination of olefin on to a sterically restricted site on a crystalline chloride of low-valent titanium, followed by a $\pi \rightarrow \sigma$ rearrangement.

The thermal decomposition of certain alkylaluminium compounds in the temperature range 180–310° can result in the deposition of a coherent and adherent film of metallic aluminium. Suitable compounds include Et_2AlH (sprayed, as a solution in hydrocarbon, on to inductively heated metal) and mixtures of Bu^i_3Al with isobutene. The cost of aluminium coatings obtained this way is likely to be competitive with that of galvanizing and tin plating, whereas the corrosion resistance of aluminium to most environments is better than that of tin or zinc. This application, which is in its early stages of development, could result in a very substantial increase in the scale of the aluminium alkyl industry.

(iii) Gallium, indium, and thallium. Most of the organic chemistry of gallium, indium, and thallium that merits inclusion in this book is included in the section below on co-ordination compounds. Here are mentioned only lower oxidation states and the instability of hydrides.

Lower oxidation states become relatively more stable with increasing atomic weight in the IIIB group. Gallium(II) and thallium(II) compounds have been obtained by reactions such as,

$$2Na + 2Me_3GaNH_3 \longrightarrow Na_2[Me_3Ga \cdot GaMe_3] + 2NH_3$$

and

$$2K + 2Me_3Tl \longrightarrow K_2[Me_3Tl \cdot TlMe_3]$$

The reaction of $K_2Ga_2Me_6$ with $MeSiCl_3$ is of some interest since all the Ga—C bonds are broken and the Ga—Ga bond remains intact,

$$K_2Ga_2Me_6 + 2MeSiCl_3 \longrightarrow 2Me_4Si + K_2Ga_2Cl_6$$

though Me_3SiCl breaks both types of bond:

$$K_2Ga_2Me_6 + 8Me_3SiCl \longrightarrow 6Me_4Si + Me_6Si_2 + 2KGaCl_4$$

Though no other organoindium(I) compounds are known, the main product from the reaction between C_5H_5Na and $InCl_3$ is C_5H_5In, the In(III) compound $(C_5H_5)_3In$ being formed only in very small yield. Cyclopentadienylindium is a polymer in the crystalline state, consisting of an infinite sandwich structure with alternating C_5H_5 groups and indium atoms. In the vapour phase both C_5H_5In and C_5H_5Tl have monomeric 'half-sandwich' structures in which the metal is just above the middle of the C_5H_5 ring but on the five-fold symmetry axis.

Cyclopentadienylthallium(I) is very easy to prepare, being precipitated when cyclopentadiene is added to aqueous thallium(I) sulphate in the presence of sodium hydroxide.

The hydrides and organohydrides of gallium are thermally much less stable than those of aluminium, and nothing is known of hydrides of indium and thallium save that attempts to prepare them lead to formation of metal.

Diethylgallium hydride resembles Et_2AlH in being associated, and in adding to olefins, but it decomposes fairly quickly above about 80°.

Co-ordination chemistry

Under this heading, complexes of boron and the third group metals are considered first because they are in general less complicated than co-ordination complexes of second group metals. The co-ordination chemistry of the organic compounds of the alkali metals is mostly of recent discovery, and is concerned almost entirely with complexes of organolithium compounds; these are considered last.

(*i*) *Boron*. Studies on the co-ordination compounds of boron (mainly its halides, alkyls and hydrides) have contributed quite substantially to our understanding of the factors which influence the stability of co-ordination compounds in general. The factors which particularly concern boron complexes are inductive and steric effects and reorganization energies. We have already remarked on the relevance of steric and electronic effects to the association of aminoboranes (p 85).

Inductive effects. The formation of a co-ordination compound between, for example, boron and nitrogen as represented by the equation

$$R_3B + NR'_3 \longrightarrow R_3B \leftarrow NR'_3$$

involves the transfer of electronic charge from nitrogen to boron, i.e. from the more electronegative to the more electropositive element, a circumstance unlikely to enhance stability. This is quite a general impediment to the formation of co-ordination compounds since donor atoms, e.g. N, P, O, S, Cl are always fairly electronegative, while acceptor atoms are usually metals or relatively electropositive non-metals.

Thus it is reasonable to suppose that the stability of a co-ordination compound will be enhanced by the extent to which

(*a*) negative charge is removed from the (electropositive) acceptor atom, and

(*b*) positive charge is removed from the (electronegative) donor atom.

For example, in a compound $R_3B \leftarrow NR'_3$, if the bonds R—B were polar in the sense $R(\delta-)$—$B(\delta+)$, the negative charge would to some extent be distributed over the three groups R by induction instead of being concentrated on the boron atom. This would occur to a greater extent as R became more electron-attracting. Consequently, if this effect were the only factor influencing the stability of co-ordination complexes, acceptor properties should decrease in the sequence $BF_3 > BCl_3 > BBr_3 > BH_3 > BMe_3$. In fact, towards a variety of donor molecules, the order of acceptor strength is $BBr_3 > BCl_3 > BF_3 \sim BH_3 > BMe_3$, showing that other factors predominate in the case of the halides. However, acceptor strengths should (and do) decrease in the sequence $BF_3 > MeBF_2 > Me_2BF > Me_3B$. Thus the complex BF_3,NMe_3 is not perceptibly dissociated below 180°; $MeBF_2,NMe_3$ is 24% dissociated at 100 mm. and 135°; Me_2BF,NMe_3 is almost wholly dissociated under these conditions, and Me_3B,NMe_3 is still less stable.

Similarly, in the series of adducts formed between ammonia or amines and Me_3B, for which dissocation data and heats of formation are listed in Table VII, the electron-releasing effect of the methyl group is seen in the increasing donor strength of the series $NH_3 < MeNH_2 < Me_2NH$. However, Me_3N is a less strong donor towards Me_3B than it should be considering only inductive effects. Inductive effects are shown by the ionization potentials of the amines, also listed in Table VII.

Table VII. *Gas phase dissociation data on Me₃B-amine compounds*

Amine	Kp (atm) at 100°C	ΔH (kcal/mole)	I.P. of amine (eV)
NH_3	4·6	13·75	10·15
$MeNH_2$	0·035	17·64	8·97
Me_2NH	0·021	19·26	8·24
Me_3N	0·472	17·62	7·82
$EtNH_2$	0·075	18·00	8·86
Et_2NH	1·22	16·31	8·01
Et_3N	Too highly dissociated to be measured		7·50
Pyridine	0·301	17·00	9·27
α-Picoline	Too highly dissociated to be measured		9·02
Quinuclidine	0·020	19·94	?

Steric effects. The figures quoted in Table VII are typical of many accurate measurements, due mainly to H.C. Brown and his collaborators, which threw much light on both inductive and steric effects. Gas-phase equilibria have the advantage that complications due to solvents, solubilities, lattice-energies etc. can be avoided. The interpretation of the anomalously low heat of formation of Me_3B,NMe_3 is that, despite the electron-releasing properties of the three methyl groups of Me_3N, as a donor towards Me_3B trimethylamine is less effective than dimethylamine because when three methyl groups are attached to nitrogen they interfere sterically with those attached to boron. Towards the proton as acceptor, that is towards a very small acceptor which will not obstruct the groups attached to nitrogen, the donor strengths of alkylamines increase throughout the series $NH_3 < RNH_2 < R_2NH < R_3N$ as has been shown by studies on the interaction between, e.g. the butylamines and acids in inert solvents. As it happens, the basic strengths of alkyl amines measured in aqueous solution change in a similar manner to their donor properties towards Me_3B, i.e. $NH_3 <$ $MeNH_2 < Me_2NH > Me_3N$ (pK_b values: NH_3, 4·75; $MeNH_2$, 3·36; Me_2NH, 3·23; Me_3N, 4·20). In aqueous solution, the alkylammonium cations are stabilized by solvation involving hydrogen bonding, which is least for the ion Me_3NH^+, and most for NH_4^+. Me_3N accordingly appears to be a weaker base than Me_2NH in aqueous solution, not for reasons of steric crowding, but because the greater solvation energy of $Me_2NH_2^+$ compared with Me_3NH^+ offsets the difference between Me_2NH and Me_3N noted in inert solvents.

Steric effects in boron complexes are more clearly shown in the ethylamine series. The strongest donor to Me_3B is $EtNH_2$, the strongest base in aqueous solution is Et_2NH, and in an inert solvent Et_3N. The inductive effect of the methyl group makes α-picoline a stronger base than pyridine, but the steric effect makes it a much weaker donor to Me_3B.

A striking demonstration of the effect of steric hindrance on donor strength is the contrast between the Me_3B adducts of triethylamine and quinuclidine. The compound with Et_3N is too unstable to allow its dissociation constant to be measured, due to steric interference between the ethyl groups and the trimethylborane molecule. In quinuclidine the carbon atoms are held back from the nitrogen and do not interfere with the trimethylborane, and the compound 3.30 has a smaller dissociation constant and a larger heat of formation than any of the others in Table VII.

$$H_2C\text{------}CH_2$$
$$HC\text{--}CH_2\text{--}CH_2\text{--}N\longrightarrow BMe_3$$
$$H_2C\text{------}CH_2$$

3.30

Reorganization energies. On co-ordination, the shape of organoborane acceptors R_3B changes from the planar unco-ordinated molecule, with bond angles $\angle RBR = 120°$, to the pyramidal co-ordinated form in which the bond angles are about $109°$. In the unco-ordinated form, the boron may be regarded as sp^2 hybridized, the third p orbital being available for π-bonding between boron and the groups R. In the co-ordinated form, however, the boron may be regarded as sp^3 hybridized; three of the sp^3 hybrid orbitals are used in σ-bonding between boron and the groups R, but the fourth is involved in the new bond to the donor atom and consequently not available for π-bonding between boron and the groups R to any great extent. Co-ordination thus results in both the loss of probably most of the π-bond energy of the molecule R_3B, and in a modification

H

in the σ-bond energy as the character of the R—B bonds changes with the hybridization of the boron. The overall energy change is referred to as the reorganization energy. Some small energy of reorganization of the donor molecule on co-ordination may also need to be taken into account, although the donor molecule does not normally undergo any major change in shape on co-ordination, although some restriction may be imposed on the conformation of bulky groups.

Attempts have been made to assess the reorganization energies of several boron compounds. Cotton and Leto used simple MO theory to estimate the π-bond energy of the planar boron halides. They suggested that the loss of this π-bond energy accounted for most of the reorganization energy of the halides, the changes in σ-bond strengths being small, and obtained the following reorganization energy values:

$$BF_3, 48\cdot3; BCl_3, 30\cdot3; BBr_3, 26\cdot2 \text{ kcal/mole}$$

These values represent maximum values in that complete absence of multiple bonding is assumed for the co-ordinated form. It is suggested that unless a reorganized BF_3 molecule forms a bond to a particular donor with the release of at least $18\cdot0$ kcal/mole *more* energy than is released when reorganized BCl_3 forms a bond to the same donor, it will be the weaker Lewis acid. Clearly, the differences between the reorganization energies of the boron halides are the reason for their relative acceptor strengths, $BF_3 < BCl_3 < BBr_3$.

Comparatively high values for the reorganization energy of a molecule BX_3 are likely when boron is bound to fluorine, oxygen or nitrogen, as the size of these atoms is such as to allow considerable π-bonding to boron. Trimethylborate is consequently a weak acceptor, and $B(NMe_2)_3$ appears devoid of acceptor properties. The size disparity between boron and the heavier elements of Groups V and VI should lead to less π-bonding, and accordingly lower reorganization energies for derivatives of these heavier elements compared with the first row elements giving group trends similar to that observed with the halogens.

Boron-carbon multiple bonding may also be significant in certain organoboranes, particularly aryl- and vinyl derivatives, in which the vacant boron p orbital offers scope for the extension of the π-electronic system, and in methyl derivatives, where the hyperconjugative effect can operate. Estimates of ca. 20 kcal/mole have been made for the reorganization energies of both vinyl$_3$B and Me$_3$B, for example.

Borane complexes. An effect similar to the hyperconjugative effect of methyl groups appears to influence the stability to dissociation of borane adducts BH_3L, which are frequently more stable than would be expected from the arguments already outlined. Thus diborane displaces BF_3 from its trimethylphosphine complex, and BH_3 alone of BX_3-type compounds

$$F_3B,PMe_3 + 1/2B_2H_6 \longrightarrow H_3B,PMe_3 + BF_3$$

forms a complex with PF_3 and CO, ligands which normally use suitable vacant orbitals for π-bonding in their complexes, the electrons for this originating on the Lewis acid. Borane is thus unique among the Lewis acids of Group III elements in having class 'B' character.

The ability of borane to act simultaneously as a σ-acceptor and a π-donor illustrates its relationship to the oxygen atom and other members of the isoelectronic series $O=$, $HN=$, $H_2C=$, and $H_3B=$. This relationship is illustrated well by some reactions of the adduct H_3B,CO which may be regarded as analogous to OCO, carbon dioxide. For example, borane-carbonyl is absorbed by alcoholic alkali, moist calcium oxide or even by $NaOH$/asbestos as used to absorb CO_2 in gas analysis, forming salts of the anion $H_3BCO_2{}^{2-}$ (cf. $CO_3{}^{2-}$). Moreover, with ammonia, methylamine or dimethylamine it gives crystalline air- and water-stable adducts $H_3B,CO,2NR_2H$ which have ionic structures $R_2NH_2{}^+[R_2N\cdot CO\cdot BH_3]^-$ in which the anions are isoelectronic with carbamates $R_2NCO_2{}^-$. The O, N, C, B skeleton of the methylamine derivative 3.31, for example, has been shown by an X-ray diffraction study to be very nearly planar. The isoelectronic relationship between $\rightarrow BH_3$ and $-CH_3$, relating the anion

3.31   3.32

$[MeHNCOBH_3]^-$ to N-methylacetamide $MeHNCOCH_3$, is also illustrated by 3.31. Trimethylamine and borane-carbonyl form an adduct believed to have the structure 3.32 (cf. Bu^tCOCH_3).

Boronium salts. Many examples of boronium salts $R_2BL_2{}^+X^-$ are known in which the boronium cation R_2B^+ (R can be H, F, Cl, Br, alkyl or aryl) is stabilized by two donor molecules L (generally amines) or a single bidentate ligand such as bipyridyl or phenanthroline. The diammoniate of diborane, $(NH_3)_2BH_2{}^+BH_4{}^-$ is such a salt. Among cationic organoboron complexes, the aryl derivatives are the most stable. For example, diphenylbipyridylboronium chloride Ph_2B bipy$^+Cl^-$ is stable to air and mild aqueous acid conditions, and is conveniently crystallized from hot water. It can be prepared from bipyridyl and Ph_2BCl, Ph_2BOH or even the ethanolamine ester of Ph_2BOH:

$$Ph_2BCl \xrightarrow{\text{bipy; } C_6H_6} Ph_2B \text{ bipy }^+Cl^- \xleftarrow{\text{bipy; aq. HCl}} Ph_2BOH \text{ or } Ph_2BOC_2H_4NH_2$$

The iodides and some other salts of the $[Ph_2B\ bipy]^+$ and $[Ph_2B$ phenan-throline$]^+$ cations are yellow or orange in colour, because of charge-transfer transitions from the anion to the co-ordinated amine of the cation; in aqueous solution they are colourless. Intra-ionic charge transfer appears to occur with salts of the phenylenedioxybipyridylboronium cation (3.33): for example the perchlorate has a yellow colour. It is suggested that electron transfer occurs from the 'lone-pairs' of the oxygen atoms of the catechol residue into the π-electron system of the bipyridyl.

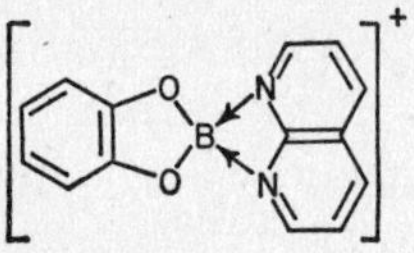

3.33

Anionic complexes. The type $M^+BR_4^-$ has already been mentioned in connection with tetraphenylborates (p 77). Tetra-alkylborates are also known, and can be prepared from e.g. $R_3B + RLi$, conveniently in benzene:

$$Me_3B + EtLi \longrightarrow LiBMe_3Et$$

The example shown is soluble in benzene, a suitable solvent for recrystal-lization, and may have a structure similar to those of analogous aluminium compounds (p 108) or the isoelectronic beryllium alkyls (p 43) with alkyl bridges linking adjacent lithium and boron atoms, a structure nevertheless unusual in view of the absence of association in the boron alkyls themselves. The tetra-alkylammonium salt (isoamyl)$_4$NB(isoamyl)$_4$ in which the cation and anion are large symmetrical ions of virtually the same size and presumably similar mobilities has proved a useful reference electrolyte for the evaluation of single ion conductivities in such solvents as $MeCN$, $MeNO_2$ and $PhNO_2$.

(ii) Aluminium and the Group IIIB elements. Towards Me_3M (M = Al, Ga, In) the order of basicity of representative Group V and VI methyls is: $Me_3N > Me_3P > Me_3As$ and $Me_2O > Me_2S > Me_2Se$. Relatively little information is available concerning trimethylthallium, which is a weaker acceptor than the other Group III trimethyls (which act as 'A' type acceptors), but the above sequence probably is true for Me_3Tl also.

The stronger bases such as Me_3N, Me_3P and Me_2O react with trimethyl-aluminium,

$$2Me_3N + Me_6Al_2 \longrightarrow 2Me_3Al,NMe_3$$

and the reaction goes to completion under conditions normally encountered because the enthalpy of the reaction

$$Me_3Al \text{ (g, monomer)} + Me_3N(g) \longrightarrow Me_3Al,NMe_3 \text{ (g)}$$

exceeds that of the reaction

$$2Me_3Al(g) \longrightarrow Me_6Al_2(g) \ \Delta H = -20 \text{ kcal/mole}$$

Complex equilibria result when the heat of co-ordination is comparable with that of dimerization, as happens for example with Me_2S and Me_2Se.

Steric factors, which have so much effect on the relative stabilities of co-ordination complexes of organoboranes, are less important for aluminium and the other Group III metals (though they can be significant when big enough groups are involved, as mentioned below). For example, the heat of formation of the Et_3Al complex of 2,6-dimethylpyridine is only about 3 kcal/mole less than that of the pyridine complex, whereas the former (sterically hindered) base does not combine with Me_3B, and pyridine forms Me_3B,py, $\Delta H = -15$ kcal/mole.

Organoaluminium compounds are not only stronger acceptors (Lewis acids) than their boron analogues, but are also much more reactive to protonic acids, probably on account of the higher polarity of aluminium-carbon bonds (more carbanion character). Trimethylaluminium forms a 1:1 complex $Me_3Al \leftarrow NHMe_2$ with dimethylamine, and the effect of the $N \rightarrow Al$ co-ordinate bond is to enhance the carbanionic character of the methyl groups and also the protonic character of the NH. It is therefore not surprising that $Me_3Al \leftarrow NHMe_2$ readily eliminates methane (about 110°):

$$2Me_3Al \leftarrow NHMe_2 = (Me_2Al \cdot NMe_2)_2 + 2CH_4$$

The hypothetical monomer, $Me_2Al–NMe_2$, would contain 3-co-ordinate, and hence co-ordinatively unsaturated aluminium and nitrogen. Moreover the acceptor strength of aluminium is increased when more electronegative groups are bound to it, i.e. the aluminium in $Me_2Al-NMe_2$ would be a stronger acceptor than aluminium in Me_3Al. Similarly, the basic character of nitrogen in $Me_2Al-NMe_2$ is greater than in Me_2NH or Me_3N.

The unsaturation in $Me_2Al-NMe_2$ might conceivably be relieved by the formation of a double bond, as in the boron analogue $Me_2B \rightleftharpoons NMe_2$, but this does not happen for reasons mentioned on p 36. Instead, this and most other similar compounds formed by second and third group elements achieve higher co-ordination numbers by the formation by each element of an additional single bond, leading to dimers, trimers or polymers.

Some of the factors which determine whether, in such circumstances, dimers, trimers or polymers are formed have already been discussed (p 87) in connection with boron.

The series $Me_2M \cdot NMe_2$ (M = Al,Ga,In) is of some interest in this respect since the members occur both as dimers and as polymers. They are dimeric in solution, and as vapour, and have four-membered ring structures such as

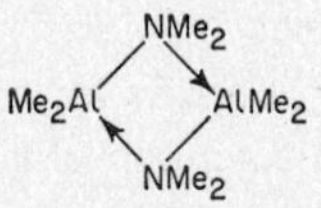

$$
\begin{array}{c}
NMe_2 \\
Me_2Al \diagup \diagdown \ AlMe_2 \\
\diagdown \diagup \\
NMe_2
\end{array}
$$

However, at temperatures in the range 50–80° all three compounds undergo a reversible transition to a glassy polymeric form (as shown by optical isotropy and X-ray diffraction). Material condensing from the vapour below the transition temperature range does so giving the crystalline form, and material condensing above this temperature forms the glass. When the glasses are heated, they do not soften gradually over a wide temperature range but have the appearance of 'melting' to a normal liquid over a range of a few degrees only. This effect must be due to a relatively rapid change with temperature of the mean degree of polymerization.

	Crystal $\rightleftharpoons$ *glass*	*Glass* $\rightleftharpoons$ *liquid*
$Me_2Al \cdot NMe_2$	50–70°	153–156°
$Me_2Ga \cdot NMe_2$	55–80°	157–161°
$Me_2In \cdot NMe_2$	70–80°	174–175°

Alcohols and thiols react with the Group III trialkyls, but alkane elimination is so fast that the intermediate co-ordination complex is not isolated. The products are associated, e.g. $(Me_2AlOMe)_3$, $(Me_2GaSMe)_2$ for reasons considered in connection with Me_2AlNMe_2.

The enhancement of the donor strength of oxygen when it is bound to an electropositive metal (as well as to carbon) as in metal alkoxides, relative to that of ether oxygen C—O—C, is well illustrated by the reaction

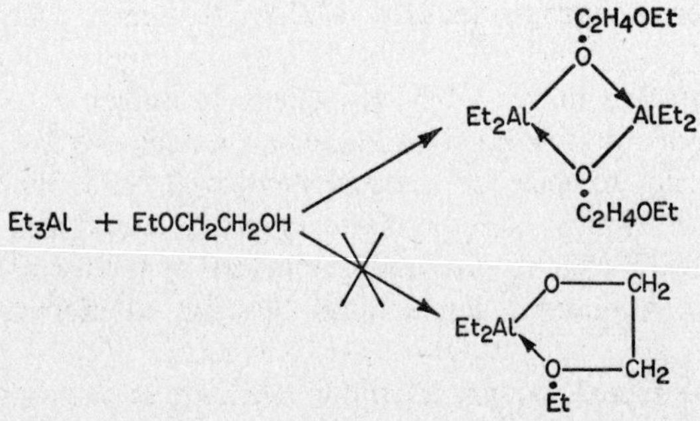

$$
Et_3Al \ + \ EtOCH_2CH_2OH
$$

$$
\begin{array}{c}
C_2H_4OEt \\
\overset{\bullet}{O} \\
Et_2Al \diagup \diagdown \ AlEt_2 \\
\diagdown \diagup \\
\overset{\bullet}{O} \\
C_2H_4OEt
\end{array}
$$

$$
\begin{array}{c}
O\!-\!\!-\!CH_2 \\
Et_2Al \diagup \qquad | \\
O\!-\!\!-\!CH_2 \\
\overset{|}{Et}
\end{array}
$$

in which the dimer is formed, even though the entropy factor favours the chelate monomer.

The co-ordination complexes of the Group III alkyls are less reactive, for example to oxygen, than the trialkyls themselves, though some of the complexes such as Me_3Al,OEt_2 spontaneously inflame in air. When two (or more) of the atoms bound to the metal are electronegative atoms like oxygen or nitrogen, then the remaining metal-carbon bonds are commonly much less reactive. The increase in carbanion character of alkyl (bound to metals) resulting from co-ordination is more than compensated by the inductive effect of the presence of two or more electronegative atoms. Since all aluminium-carbon compounds are thermodynamically unstable to hydrolysis and oxidation, the effects discussed above are kinetic. Steric influences are often substantial when large groups like *t*-butoxy and triphenylmethoxy are bound to the metal, a striking example being $(Me_2AlOCPh_3)_2$ which does not dissolve even in dilute hydrochloric acid as nucleophilic attack at the metal is sufficiently impeded.

Whereas aluminium alkyls react with most hydroxy compounds (unless there are steric complications) with loss of all three alkyl groups, such reactions tend to stop after one or two alkyl groups have been eliminated from gallium and indium alkyls. Thallium trialkyls are hydrolysed only as far as R_2TlOH, and to R_2Tl^+ cations in acid solution (these are considered later).

Unsaturated basic groups such as cyanide and carbonyl could form co-ordination complexes with R_3M but could also react by addition of R—M across the multiple bond. For example, trimethylaluminium forms a complex $Me_3Al \leftarrow N \vdots CPh$ with phenyl cyanide, but when this is heated it re-arranges by addition of methyl-aluminium to $- C \vdots N$ giving $(Me_2AlN \vdots CMePh)_2$, itself hydrolysed to acetophenone. A similar complex prepared from Me_3Al and Bu^tCN has had its cyclic structure confirmed by X-ray analysis:

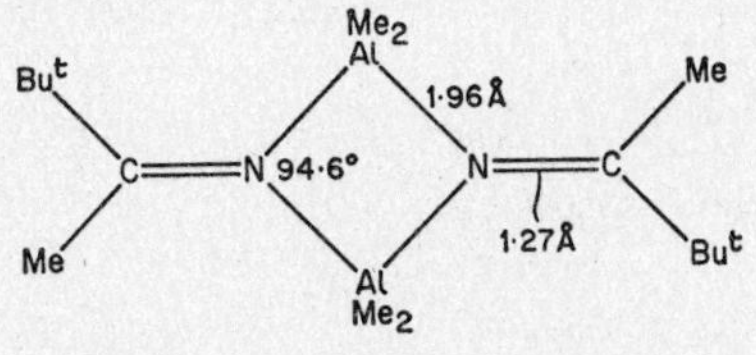

The addition of Al—H to unsaturated bonds takes place much more readily than that of Al—C, and by way of example Me_2AlH reacts with cyanides forming aldimine derivatives $(Me_2AlN \vdots CHR)_2$.

Reactions between aluminium alkyls and carbonyl compounds may be complicated by enolization and reduction processes, and the simple adducts $R'_2CO \rightarrow AlR_3$ are too labile to be isolated. In one typical experiment the reaction between Et_3Al and Et_2CO in 1:1 molar ratio (25°) gave these products in relative amounts shown:

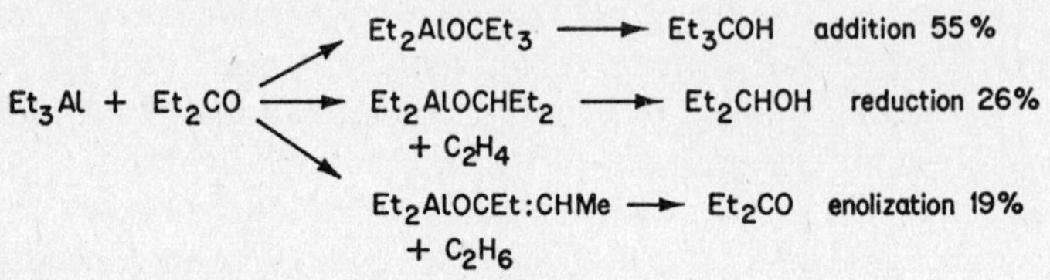

$$Et_3Al + Et_2CO \longrightarrow \begin{cases} Et_2AlOCEt_3 \longrightarrow Et_3COH \quad \text{addition } 55\% \\ Et_2AlOCHEt_2 \longrightarrow Et_2CHOH \quad \text{reduction } 26\% \\ \quad + C_2H_4 \\ Et_2AlOCEt:CHMe \longrightarrow Et_2CO \quad \text{enolization } 19\% \\ \quad + C_2H_6 \end{cases}$$

The reduction process involves hydrogen transfer from β-carbon

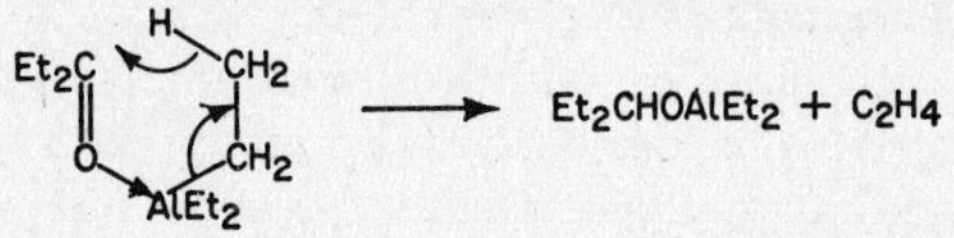

$$Et_2C \overset{\displaystyle H}{\underset{\displaystyle O}{\diagdown}} \overset{CH_2}{\underset{CH_2}{\diagup}} \overset{}{\underset{AlEt_2}{}} \longrightarrow Et_2CHOAlEt_2 + C_2H_4$$

and does not occur with trimethyl- or triphenyl-aluminium. Hydrogen transfer can also take place by another mechanism when the carbonyl: Al ratio is greater than unity. Thus, the phenylation of benzaldehyde follows the course,

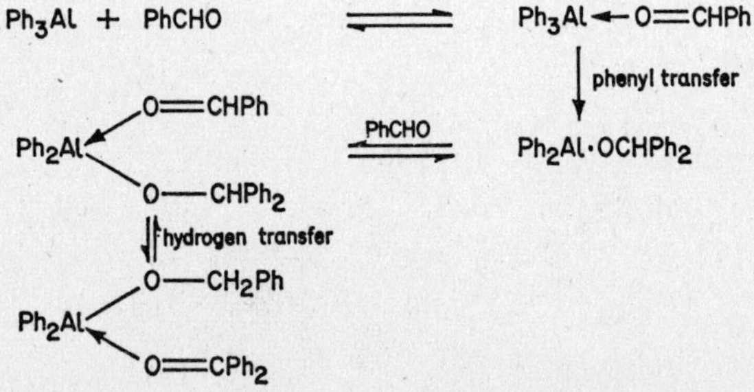

$$Ph_3Al + PhCHO \rightleftharpoons Ph_3Al \leftarrow O{=}CHPh$$

$$\downarrow \text{phenyl transfer}$$

$$Ph_2Al{\diagdown} \overset{O{=}CHPh}{\underset{O{-}CHPh_2}{}} \quad \overset{PhCHO}{\rightleftharpoons} \quad Ph_2Al{\cdot}OCHPh_2$$

$$\updownarrow \text{hydrogen transfer}$$

$$Ph_2Al{\diagdown} \overset{O{-}CH_2Ph}{\underset{O{=}CPh_2}{}}$$

and hydrolysis gives about equal amounts of benzyl alcohol, diphenyl-carbinol and benzophenone (compare the Meerwein-Pondorff reduction).

The organoaluminium halides are all co-ordination complexes, the chlorides, bromides and iodides being dimeric because of the tendency of these halogens to form two bonds roughly at 90°. Methylaluminium dichloride is quoted here as sole example, as its crystal structure has been determined:

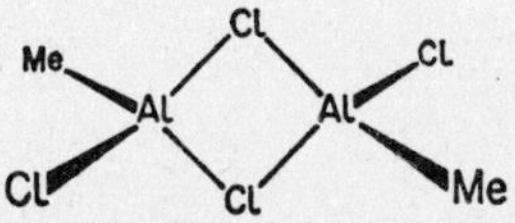

The structure of $Me_4Al_2Cl_2$, which also has an $AlCl_2Al$ bridge, has been determined by electron diffraction and by vibrational spectroscopy.

The fluorides differ, mainly because two-co-ordinate fluorine seems to prefer much larger bond angles (up to 180°, see p 110). This leads to oligomer and polymer formation. Both dimethyl- and diethylaluminium fluorides, though volatile (bp 68–70°/15 mm and 90–91°/0·4 mm), are very viscous glass-like liquids. The higher members R_2AlF are mobile and not so associated. Diethylaluminium fluoride is tetrameric in benzene, this being compatible with 2-co-ordinate fluorine forming co-linear or nearly co-linear bonds (cf. $[Me_2AlCN]_4$ which also has a linear co-ordination group). The fluorine-bonding is so strong that the tetramer is depolymerized neither by ethers nor by trimethylamine, unlike $Me_4Al_2Cl_2$ or $(AlCl_3)_x$.

The organohalides of gallium and indium resemble those of aluminium, but have been relatively little studied. Little is known of the fluorides. The thallium compounds, R_2TlX, in contrast have been known for a long time and many more of these have been described than of R_3Tl. This is because reaction of the commonest laboratory alkylating agent, RMgX in ether, with $TlCl_3$ gives R_2TlX (and often much TlX as well). These compounds (R_2TlX) resist hydrolysis, and those that dissolve in polar solvents are electrolytes, e.g. Me_2TlF, Me_2TlNO_3. The crystal structure of Me_2TlI shows the presence of linear [Me—Tl—Me] units, probably as cations iso-electronic with Me—Hg—Me, and each thallium has four iodine atoms (at an ionic distance) in a square making the thallium six-co-ordinate.

Most R_2Tl^+ cations act as rather weak acceptors. For example, Me_2TlCl dissolves in concentrated aqueous ammonia but is precipitated as ammonia evaporates. Dimethylthallium perchlorate forms several complexes with nitrogen bases, and the Me—Tl—Me angle becomes less than 180° (it is 168° in Me_2Tl phenan ClO_4). The acceptor strength of thallium in R_2Tl units is considerably increased when R has strong electron-withdrawing character, and in $(C_6F_5)_2TlBr$ it is enough to cause dimerization like the aluminium compounds discussed earlier:

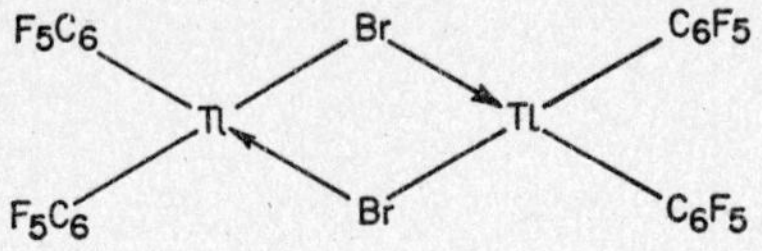

Many other examples are known of four-co-ordinate bispentafluorophenyl-thallium compounds, and there are a few instances, e.g. $(C_6F_5)_2 (Ph_3AsO)_2 TlNO_3$, of five-co-ordinate thallium. These high co-ordination numbers obtain only when the metal is bound to groups which are at the same time electron-attracting and not very polarizable, since this combination results in excessive net positive charge on the metal conflicting with the Pauling electroneutrality principle and requiring co-ordination with a base for its reduction to an acceptable value (compare aluminium: six-co-ordination found with F and O, Cl borderline, and four-co-ordination with Br, I, C, S etc.).

Some of the anionic complexes, particularly of aluminium are both of structural and synthetic interest. Sodium reacts with triethyl- but not with trimethyl-aluminium:

$$3Na + 4Et_3Al \longrightarrow 3NaAlEt_4 + Al$$

A range of such complexes is known, and many of the lithium and sodium compounds are soluble in hydrocarbons. The lithium compound, $LiAlEt_4$, can be sublimed and in the crystalline state has a polymeric structure resembling that of dimethylberyllium; each lithium and each aluminium is tetrahedrally surrounded by CH_2 groups.

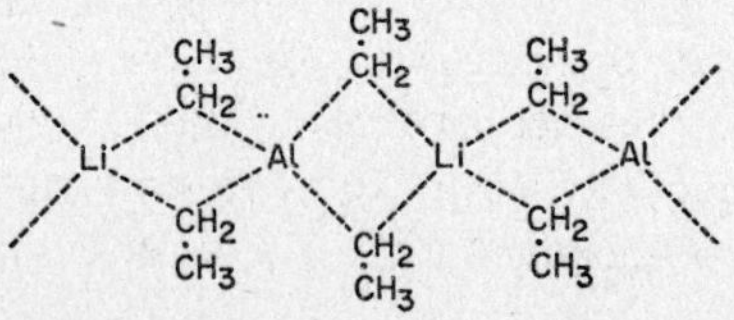

The aggregates which must be present in solutions of $LiAlR_4$ and $NaAlR_4$ in hydrocarbons and in the vapour of $LiAlEt_4$ presumably contain units in which Al—C bonds are polarized by the alkali metal to the point at which the bonding may more appropriately be described as electron-deficient, (as in crystalline $LiAlEt_4$):

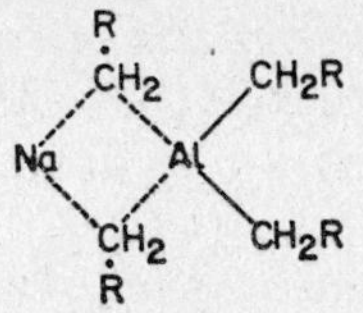

Similar compounds are formed by Group II metals, $M(AlEt_4)_2$, $M =$ Mg, Ca, Sr, Ba. Fused $NaAlEt_4$ is moderately conducting so it must be extensively ionized (sp. conductivity 0·06 at 130°), and its electrolysis gives sodium at the cathode and $Et_3Al + C_2H_4 + C_2H_6$ (the last two from the decomposition of ethyl radicals) at the anode. Anodes made of metals which form ethyl derivatives are attacked, and, for example, lead gives Et_4Pb.

These salts could be regarded as containing complexes between Et_3Al and ethyl anions, and they can be made from Et_3Al and ethyl-sodium or -potassium. Aluminium trialkyls also form anionic complexes with a range of other anions, notably hydride and halide ions. The decrease in bond strength M—A with increasing atomic weight of A, discussed in Chapter 1 (p 4), operates here also in the sense that the reaction

$$R_3Al + X^- \longrightarrow [R_3AlX]^- + Q$$

becomes less exothermic as the size of the ion X^- increases. Since the lattice energy of salts MX is bound to be much bigger than that of $M[R_3AlX]$, the former is also a relevant factor and involves the sizes both of M^+ and X^-. Thus the formation of complexes is favoured when the cation is large and the anion is small, as shown in Table VIII in which the existence of a complex is shown by +.

Substitution of alkyl by halogen also favours complex formation by increasing the Lewis acidity of the metal. Salt, for example, forms a complex with $MeAlCl_2$ but not with Me_2AlCl, and this permits the easy preparation of the latter from $Me_3Al_2Cl_3$ (p 93).

Table VIII. *Stability of* 1 : 1 *MX, Et_3Al complexes*

	H^-	F^-	Cl^-	Br^-	I^-
Li^+	+	−	−	−	−
Na^+	+	+	−	−	−
K^+	+	+	+	−	−
Rb^+	(+)	+	+	+	−
Cs^+	(+)	+	+	+	−
Et_4N^+	(+)	(+)	+	+	+

(+) Formation of these not investigated.

The hydride and fluoride complexes are of particular interest because two series are formed, e.g. $LiAlEt_3H$ and $LiAl_2Et_6H$. The fluoride complex KAl_2Et_6F contains a linear Al---F---Al system:

$$\overset{1 \cdot 80 \text{Å} \quad 1 \cdot 80}{[Et_3Al \text{ - - - } F \text{ - - - } AlEt_3]^-}$$

Since this compound evolves Et_3Al when heated,

$$KAl_2Et_6F \longrightarrow KAlEt_3F + Et_3Al$$

and since it may be obtained by carrying out the reverse reaction in ether solution,

$$KAlEt_3F + Et_3Al \cdot OEt_2 \longrightarrow KAl_2Et_6F + Et_2O$$

this provides a method for obtaining Et_3Al from its ether complex.

(*iii*) *Group II metals*. Complexes between metal alkyls and bases not containing acidic hydrogen are considered first, because elimination of hydrocarbon from complexes such as $R_2M \leftarrow NHR'_2$ (giving $RMNR'_2$) leads to various complications.

In the series, R_2M, there is a general inverse correlation between the electronegativity of the metal and the Lewis acid character of its alkyls. Basically, this is because metal–carbon bonds are strongly polarized $M(\delta+)$—$C(\delta-)$ when the electronegativity difference is big, as when M is Mg, and the metal can then accept negative charge from donor atoms without contravening the Pauling electroneutrality principle, and therefore readily forms complexes R_2ML or R_2ML_2. When the electronegativity difference is smaller, as in Me_2Hg, the effect of co-ordination would be to place unacceptably high negative charge on the metal. This effect is more pronounced with mercury than with boron, partly because in the latter case there are three carbon atoms to help take away negative charge. Other factors, such as the ns—np energy difference which is considerably greater for mercury than for zinc and cadmium and thus impedes the expansion of the covalency of mercury, are relevant but the electronegativity effect is dominant. In the series of dimethyls of Zn, Cd, Hg (Pauling electronegativities $1 \cdot 6$, $1 \cdot 7$, $1 \cdot 9$) Me_2Zn forms well-defined complexes with some tertiary amines and 2,2'-bipyridyl, Me_2Cd forms such an unstable complex with bipyridyl that it can be evaporated away at room temperature leaving a residue of bipyridyl, and Me_2Hg forms no complex with bipyridyl. If, however, the methyl groups in Me_2Hg are replaced by electron-attracting groups such as CF_3, CCl_3 or C_6F_5, then complexes such as $(CF_3)_2Hg$ bipy and $(C_6F_5)_2Hg(Ph_2PC_2H_4PPh_2)$ are formed. No compounds containing the cations RZn^+, RCd^+ or RHg^+ have been prepared, but numerous co-ordination complexes of RHg^+ are known, their interest being that the ion complexes with only *one* mole of base, and that it behaves as a 'B' type acceptor (relatively low $5d$–$6s$ separation). Aqueous solutions of methylmercury salts containing no species of greater donor strength (to mercury) than water consist of hydrated ions $[Me—Hg{\leftarrow}OH_2]^+$ together with the appropriate anions. Water, having weak 'B' donor properties, is easily displaced by ligands such as halides, sulphides or phosphines. Some of these compounds may be isolated as crystalline salts, e.g. $[MeHg{\leftarrow}PEt_3]Br$, but others disproportionate readily. For example, addition of Ph_3P to an acetone solution of m-tolylmercury chloride results in an immediate and large increase in electrical conductivity followed by a steady fall as disproportionation takes place:

$$m\text{-MeC}_6\text{H}_4\text{HgCl} + \text{PPh}_3 \longrightarrow [m\text{-MeC}_6\text{H}_4\text{Hg} \leftarrow \text{PPh}_3]^+\text{Cl}^- \longrightarrow$$
$$(m\text{-MeC}_6\text{H}_4)_2\text{Hg} + (\text{PPh}_3)_2\text{HgCl}_2 \downarrow$$

This is the basis for a useful method for preparing R_2Hg from $RHgX$ by reaction with 'B' type ligands such as cyanide or iodide as well as PR_3:

$$2\text{RHgX} + 4\text{I}^- \longrightarrow \text{R}_2\text{Hg} + \text{HgI}_4{}^{2-} + 2\text{X}^-$$

Reactions between the lighter second group dialkyls, which are 'A' type acceptors, and bases such as ethers and tertiary amines may be complicated both by steric factors and by heats of co-ordination sometimes being similar to heats of depolymerization of electron-deficient polymers.

Steric effects are more evident in the co-ordination chemistry of beryllium than in that of magnesium alkyls, because of its smaller covalent radius (compare Mg—C, $2 \cdot 16$ Å in $EtMgBr2Et_2O$, and Be—C, $1 \cdot 699$ Å, in Bu^t_2Be). Dimethylberyllium forms a stable trimethylamine complex, Me_2BeNMe_3, but although this is a compound containing only 3-coordinate metal and thus has a vacant low-lying $2p$ orbital, addition of a second molecule of amine (almost as bulky as the t-butyl group) gives an unstable adduct, $Me_2Be(NMe_3)_2$, whose dissociation pressure is 23 mm at $0°$. The dissociation

$$Me_2Be(NMe_3)_2 \; cryst \longrightarrow Me_2BeNMe_3 \; cryst + Me_3N \; gas$$

is accompanied by a considerable entropy increase, but the detachment of one dimethylamino group from the chelate complex 3.34 would not increase the number of independent particles and would not be accompanied by any significant entropy change.

3.34   3.35

It is not possible to pack the equivalent of four tertiary-butyl groups within bonding distance of one beryllium atom, so 3.35 is a non-existent compound in the form shown. A study of the temperature dependence of its ^{1}H n.m.r. spectrum shows that at room temperature only one NMe$_2$ group is complexed at one time, but that complexed and free amino groups change places rapidly: as the temperature is reduced so does the exchange rate and a distinction becomes apparent between free and complexed amino groups. In contrast, the ^{1}H n.m.r. spectrum of the magnesium analogue of 3.35 shows no temperature dependence, since magnesium is big enough to accommodate all four groups at the same time.

Electron-deficient polymers such as $(Me_2Be)_x$ and $(Me_2Mg)_x$ react only with strong bases that release enough energy forming donor-acceptor bonds to compensate for the energy needed to break down the electron-deficient polymer. Dimethylberyllium, which requires about 20 kcal/gram atom of beryllium to break it down to monomer units, is completely depolymerized by reaction with typical tertiary amines or pyridine. With weaker bases such as Me$_2$O, Et$_2$O, Et$_2$S and Me$_3$P equilibria still involving methyl-bridged species can obtain,

In dilute solution in diethyl ether, Me$_2$Be is monomeric, the base then being in great excess.

The tendency for methyl groups to form electron-deficient bridges between magnesium atoms is great, and dimethylmagnesium has a degree of association between one and two in diethyl ether, indicating the existence of equilibria such as that written above for the Me$_2$Be/PMe$_3$ reaction. Further, it crystallizes from ether at room temperature without complexed ether, i.e. as Me$_2$Mg polymer. Other alkyl groups have somewhat weaker tendencies to form electron-deficient bridges, and some of these form distinct ether complexes. Magnesium dialkyls form well-defined monomeric complexes with tetramethylethylenediamine, analogous to 3.34.

The second group alkyls form complexes with heterocyclic bases many of which are brightly coloured, for example 2,2'-bipyridyl forms yellow Me$_2$Be bipy, red Et$_2$Be bipy and orange-red Et$_2$Zn bipy. Similarly, o-phenanthroline forms violet coloured complexes with magnesium dialkyls. The colours of complexes formed by heterocyclic bases are often due to transitions in which an electron is transferred into the lowest vacant π orbital of the heterocyclic system. The familiar red colour of the ferrous-bipyridyl ion is due to such processes, in which an electron is transferred from the Fe^{++} giving, in the excited state, Fe^{+++}. In the zinc alkyl complexes, electron transfer could possibly be from the filled $3d$ shell of the zinc, but in the beryllium and magnesium complexes the most likely electron source is the pair of metal-carbon bonds, each bond becoming of order 0·75 in the excited state.

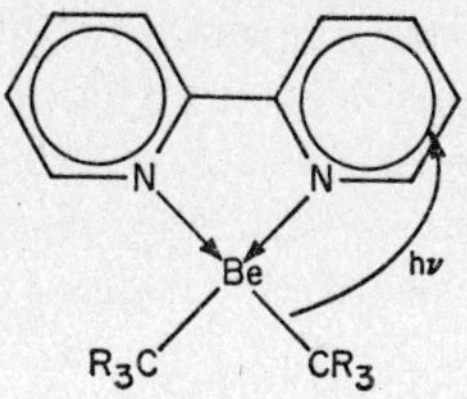

The absorption moves to the red when electron-releasing, and to the blue when electron-attracting groups are bound to the metal (e.g. Ph_2Be bipy is pale yellow, $[p\text{-}Me_2NC_6H_4]_2Be$ bipy is orange).

Secondary amines readily eliminate hydrocarbon on reaction with alkyls R_2M, though the intermediate can sometimes be isolated:

$$Me_2Be + Me_2NH \longrightarrow Me_2Be,NHMe_2 \longrightarrow MeBeNMe_2 + CH_4$$

In this instance the crystalline dimethylamine complex decomposes as it melts (44°). For the reasons already discussed in connection with the analogous aluminium compound Me_2AlNMe_2, dimethylaminomethyl-beryllium associates. Up to 50° it is a poly- or oligo-meric glass then it 'melts' between 51 and 54°. As vapour or in solution it is a cyclic trimer 3.36.

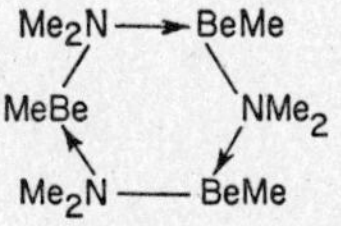

3.36 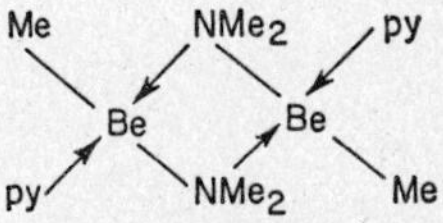   3.37

In 3.36 it will be noted that though the nitrogen atoms are 4-co-ordinate, the metal atoms are only 3-co-ordinate. Such compounds therefore have a vacant orbital on each metal atom, and this allows reaction with further molecules of base (if steric conditions permit). Thus 3.36 adds either one molecule of pyridine for each beryllium atom, giving 3.37, or two molecules giving monomeric $Me(Me_2N)Be\,py_2$. The amino alkyls $(RMNR'_2)_n$ of the second group elements, and those $(R_2MNR'_2)_n$ of the third group metals are labile relative to those of boron, and the extent of their association is determined by thermodynamic rather than by kinetic factors. Steric effects predominate. Cyclic trimers like 3.36 would be preferred to dimers containing a four-membered Be_2N_2 ring as in 3.37 on account of the valency angle distortion from 109° to an average of 90° in the dimers. However, as the size of R and R′ is increased, steric interference between them becomes serious in trimers before it does in the corresponding dimer. Hence these compounds dissolve as trimers when R and particularly R′ are small, and as dimers otherwise. For example, all the dimethyl-aminoberyllium alkyls so far known are trimers, viz. $(MeBeNMe_2)_3$, $(EtBeNMe_2)_3$, $(Pr^iBeNMe_2)_3$ and $(PhBeNMe_2)_3$, whereas larger substituents giving dimers are illustrated by $(MeBeNPr^n_2)_2$, $(MeBeNPh_2)_2$ and $(PhBeNPh_2)_2$. The strong tendency for alkyl groups bound to magnesium to form bridges between magnesium atoms has already been mentioned, and can result in the association of $RMgNR′_2$ forming polymers. For example, Pr^i_2Mg eliminates C_3H_8 on reaction with Ph_2NH in ether, giving $Pr^i(Ph_2N)Mg(OEt_2)_2$ which readily loses ether forming an insoluble polymer in which polymerization would be propagated both by the nitrogen atoms and by the isopropyl groups

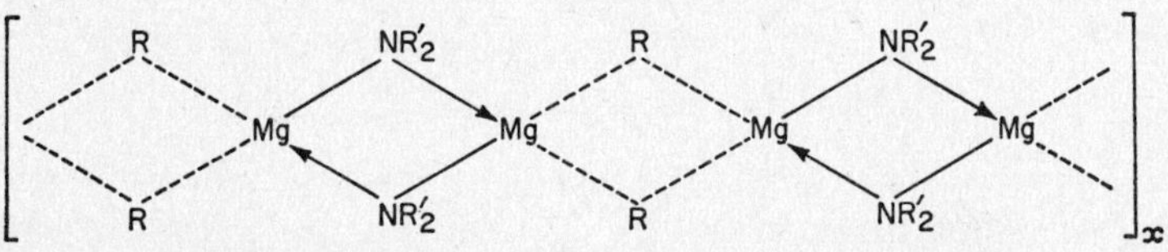

Such association can be prevented when the groups R′ are bulky, for example Pr^i_2Mg with Pr^i_2NH gives $(Pr^iMgNPr^i_2)_2$. Another amino-magnesium alkyl containing 3-co-ordinate metal is formed by the addition of Et_2Mg to benzylideneaniline:

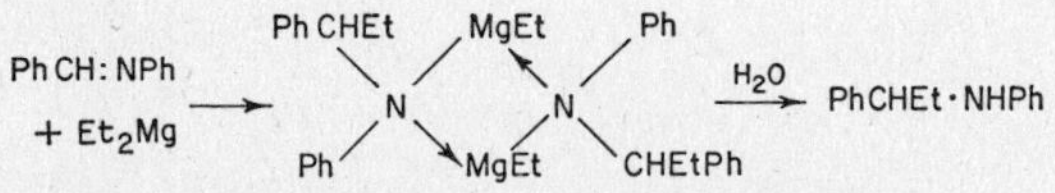

Reaction between dimethylzinc and diphenylamine yields a product $(MeZnNPh_2)_2$ in which the presence of 3-co-ordinate zinc has been shown by X-ray diffraction (Figure 21).

The Zn–N bond in this and related compounds has the interesting property of readily adding across three-atom unsaturated groups, e.g.

$$EtZnNEt_2 + CO_2 \longrightarrow EtZnO \cdot CO \cdot NEt_2$$

$$EtZnNPh_2 + PhNCO \longrightarrow EtZnNPh \cdot CO \cdot NPh_2$$

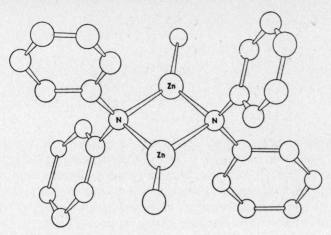

Figure 21. Crystal structure of [MeZnNPh$_2$]$_2$. (After H. M. M. Shearer and C. B. Spencer, personal communication). All the angles in the four-membered ring are right angles.

The elimination of hydrocarbon by reactions of R$_2$M and an alcohol gives products which if they were monomeric, would have two donor sites (oxygen lone-pairs) and at least two acceptor sites present in the same unit:

$$\text{RM}\!\longrightarrow\!\overset{\uparrow}{\underset{\downarrow}{\text{OR}'}}$$

Such substances, of course, associate but any kind of open ring such as

$$\underset{\text{OR}'}{\overset{\text{OR}'}{\text{RM}}}\diamond\text{MR} \quad \text{or} \quad \underset{\text{R}'\text{O}\!\longrightarrow\!\text{MR}}{\overset{\text{R}'\text{O}\!\longrightarrow\!\text{MR}}{\text{RM}}}\text{OR}'$$

in which only *one* of the lone-pairs in each oxygen atom is used would still leave the metal 3-co-ordinate. A few instances of trimeric alkylmetal alkoxides are known, e.g. (EtBeOCEt$_3$)$_3$, (Bu^{n}MgOPr^i)$_3$, (EtZnOCHPh$_2$)$_3$ and (PhHgOMe)$_3$, but tetramers are more frequently encountered. Tetrameric alkyl alkoxides include (MeBeOR)$_4$, R = Me, Et, Pr^n, Pr^i, Bu^t; (EtMgOBu^t)$_4$, (Pr^{i}MgOPr^i)$_4$, (MeZnOMe)$_4$, (MeZnOBu^t)$_4$, (MeCdOPr^i)$_4$. A tetramer is the lowest degree of association allowing both metal and oxygen to be 4-co-ordinate. The highly symmetrical structure of the methylzinc methoxide tetramer has been determined by X-ray diffraction (Figure 22).

H

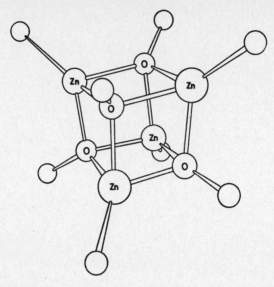

Figure 22. Crystal structure of methylzinc methoxide tetramer. (After H. M. M. Shearer and C. B. Spencer, *Chem. Comm.*, 1966, 194).

Methylzinc methoxide itself disproportionates slowly in solution, forming free dimethylzinc and a soluble methoxy-rich species, but $(MeZnOPr^i)_4$ and $(MeZnOBu^t)_4$ are free from this complication.

Inspection of Figure 22 shows that the alkoxy groups R′ in a tetramer $(RMOR')_4$ must be in a *cis* configuration relative to one another. If these groups, and R, are large enough then steric interference prevents the formation of tetramers, and dimers result in which the groups R′ can take up a planar or a *trans* configuration. For example, whereas methylberyllium *t*-butoxide (which is accessible by several routes) is a tetramer, the *t*-butyl *t*-butoxide is a dimer:

$$Me_2Be + Bu^tOH$$

$$\downarrow$$

$$Me_2Be + Me_2CO \longrightarrow (MeBeOBu^t)_4 \underset{Me_2Be}{\overset{Me_2CO}{\rightleftharpoons}} [(Bu^tO)_2Be]_3$$

$$\uparrow$$

$$Me_2Be + Bu^tO \cdot OBu^t$$

$$Bu^t_2Be + Bu^tOH \longrightarrow (Bu^tBeOBu^t)_2$$

The formation of two co-ordinate bonds by oxygen is not common, and it would not be surprising if, in such compounds, the second lone-pair were fairly easily displaced by other bases. As an example, the tetramer $(MeZnOPh)_4$ is cleaved by pyridine:

$$\tfrac{1}{2}(MeZnOPh)_4 + 2py \longrightarrow$$

Similarly, typical alkylberyllium alkoxides react reversibly with ether

$$\tfrac{1}{2}(RBeOR')_4 + 2Et_2O \rightleftharpoons$$

and, for example, $MeBeOBu^t$ is nearly tetrameric in concentrated and is dimeric in dilute solution in ether.

Alkylmagnesium alkoxides deserve some comment, particularly as they feature as intermediates in reactions between Grignard reagents and carbonyl compounds (see p 60). Those that have been examined in ether solution have degrees of association ranging from two to four. Both $MeMgOBu^t$ and $EtMgOEt$ are tetrameric even in dilute solution, though the more sterically hindered $EtMgOCEt_3$ is present as dimer.

Though, as mentioned earlier, four is the lowest degree of association of $RMOR'$ allowing both metal and oxygen to become 4-co-ordinate (in absence of donor solvents), higher degrees of association are also possible but would be less favoured by the entropy factor. Several alkyl-magnesium alkoxides are known which are 6–8 fold associated.

Reactions between R_2M and thiols $R'SH$ lead also to associated species $(RMSR')_n$. Though many of these are tetramers, e.g. $(EtBeSEt)_4$, $(EtMgSBu^t)_4$, higher degrees of association sometimes occur, one of the more interesting examples being the pentameric $(RZnSBu^t)_5$, $R = Me$ or Et. The methyl compound, whose structure has been determined, contains 3-, 4- and 5-co-ordinate sulphur, the environment of the latter being unique to this compound (Figure 23). As expected, all the zinc atoms are 4-co-ordinate. The isopropyl compound $MeZnSPr^i$ is an octamer in the crystalline state, with a symmetrical cage structure.

(*iv*) *The alkali metals.* The series R_3B,L and R_2BeL_2 (in which L is a base such as ether or tertiary amine) is not completed by $RLiL_3$, still less are complexes $RNaL_3$ known. There is no evidence for the existence of any complexes of organo-sodium or -potassium compounds other than those which have, beyond reasonable doubt, ionic constitutions and which

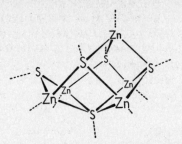

Figure 23. Structure of $(MeZnSBu^t)_5$. Carbon atoms are omitted but dotted lines show the directions in which alkyl groups lie. (G. W. Adamson, H. M. M. Shearer and C. B. Spencer, *Acta Cryst.*, 1966, **21**, A135).

therefore contain alkali metal ions co-ordinated to bases in the same sense as they are solvated by water in aqueous solutions.

The lithium atom in hypothetical complexes $RLiL_3$ would be carrying too high a negative charge, and the molecule would be in conflict with the electroneutrality principle. The tetramethylethylenediamine (TMED) complexes of butyl-lithium and some other organolithium compounds have already been mentioned in connection with metalation reactions (p 49). The effect of forming two Li—N co-ordinate links is to make the carbon atom bound to lithium very strongly nucleophilic.

There is one respect in which the formation of complexes of RLi resembles that of R_2Mg complexes (an example of the 'diagonal relationship') Alkyl groups form such strong electron-deficient bridges between magnesium atoms that these can co-exist with donor molecules. Examples such as the degree of association of Me_2Mg being between 1 and 2 in ether have been discussed earlier. Similarly, there is evidence that electron-deficient alkyl bridges between lithium atoms are broken down only when one or more of these conditions are met:

(a) the organic group R forms a particularly stable anion,
(b) the alkyl, RLi, reacts with a strong chelating base.

Compounds such as $PhCH_2Li$ and Ph_3CLi dissolve in ether forming solvated lithium cations and stable benzyl or benzyl-type anions (also solvated). Benzyl-lithium has been shown to be monomeric in tetrahydrofuran. The second condition, (b), is met in the TMED complexes.

Most lithium alkyls retain some electron-deficient $(RLi)_n$ aggregate in ethers, some observed degrees of association being given in Table IX.

Triethylenediamine, which is not a chelating base, forms a range of complexes with lithium alkyls. These complexes have the composition $(RLi)_2N(C_2H_4)_3N(RLi)_2$, but differ sharply from the TMED complexes in their sparing solubility in hydrocarbons, the TMED complexes being very

Table IX. *Association of Organolithium Compounds*
Degree of association in:

	Hydrocarbon	Diethyl Ether	Tetrahydrofuran
MeLi	insol.	4	4
EtLi	6	4	(a)
Bu^nLi	6	4	4
Bu^tLi	4	(a)	(a)
PhLi	insol.	2	2
$PhCH_2Li$	(a)	(a)	1

(a): not known.

soluble. It has been very reasonably suggested (T. L. Brown) that in the crystalline state the triethylenediamine complexes are polymeric, and retain the $(RLi)_4$ units:

$$-N \to (RLi)_4 \leftarrow N(C_2H_4)_3N \to (RLi)_4 \leftarrow N(C_2H_4)_3N \to (RLi)_4 \leftarrow N-$$

These complexes have strong carbanionic activity, like RLi(TMED), and crystal structures will be awaited with interest.

There is evidence also that oxygen in lithium alkoxides can complex with lithium alkyls. Both the strong donor strength of alkoxy oxygen and the fact that lithium n-butoxide is soluble in hydrocarbon solutions of n-butyl-lithium have been mentioned earlier.

BIBLIOGRAPHY

Organolithium compounds
(a) Structures and constitution; T. L. Brown, *Advances in Organometallic Chemistry* **3**, 1965, 365; (b) Analysis; R. L. Eppley and J. A. Dixon, *J. Organometallic Chem.* **8**, 1967, 176; S. C. Watson and J. F. Eastham, *ibid.* **9**, 1967, 165.

Hydrocarbon anions
E. de Boer, *Advances in Organometallic Chemistry* **2**, 1964, 115.

Bonding in electron-deficient molecules, including a molecular orbital description of the trimethylaluminium dimer
R. E. Rundle, *Survey of Progress in Chemistry* **1**, 1963, 81

Grignard reagents, constitution and mechanism of their reactions with ketones
E. C. Ashby, *Quart. Rev.* **21**, 1967, 259

Organo-aluminium chemistry

K. Ziegler, 'Organometallic Chemistry', edited by H. Zeiss (Reinhold, New York, 1960) Chapter 5; R. Köster and P. Binger, *Advances in Inorganic Chemistry and Radiochemistry* **7**, 1965, 263

Stereoregular olefin polymerization

C. E. H. Bawn and A. Ledwith, *Quart. Rev.* **16**, 1962, 361.

K. Ziegler, Advances in Organometallic Chemistry, **6**, 1968, 1. A fascinating account of the author's important contributions to this subject during forty years research.

Organoboron compounds

M. F. Lappert, Chapter 8 of 'The Chemistry of Boron and Its Compounds'. (Wiley, New York, 1967). Other articles in this book and in 'Progress in Boron Chemistry', edited by H. Steinberg and A. L. McCloskey (Pergamon, London) Volume 1, 1964, survey a range of topics in organoboron chemistry.

Borazines

J. C. Sheldon and B. C. Smith, *Quart. Rev.* **14**, 1960, 200; E. K. Mellon and J. J Lagowski, *Advances in Inorganic Chemistry and Radiochemistry*, edited by H. J· Eme1éus and A. G. Sharpe, **5**, 1963, 259.

Carboranes

E. L. Muetterties and W. H. Knoth, 'Polyhedral Boranes' (Edward Arnold and Marcel Dekker, London and New York, 1968).

M. F. Hawthorne, Plenary Lecture at Third International Symposium on Organometallic Chemistry, (Butterworths, London, 1968), *Pure and Applied Chemistry*, **17**, No. 2. 1968. Accounts of Chemical Research, **1**, 1968, 281. Transition metal complexes derived from carboranes.

CHAPTER FOUR

Organometallic compounds of elements of main groups IV and V

Introduction

In this chapter we shall consider the B subgroup elements silicon, germanium, tin, lead, arsenic, antimony and bismuth. The organic chemistry of these elements, notably of silicon, tin and arsenic and to a lesser extent of germanium and antimony is particularly extensive, and only a few salient features will be outlined here.

Generally, these elements differ from those discussed in Chapter 3 in the following important respects; their bonds to carbon are appreciably less polar in the sense $M(\delta+)$—$C(\delta-)$, and their simple organic derivatives, having no low-energy vacant molecular orbitals, do not normally function as acceptor molecules and are not associated through electron-deficient alkyl or aryl bridges. They are accordingly less reactive to nucleophiles; most are unaffected by water, and many (particularly Group IV derivatives) are air-stable, although reactivity and thermal instability increase with atomic number, i.e. with increasing length and decreasing strength of the M—C bonds.

Several methods of attaching organic groups to elements of Groups IV and V have already been described in Chapter 2. The methods most generally useful in the laboratory are those using derivatives of the more electropositive metals, e.g. Grignard, organo-lithium or -aluminium reagents. Routes to industrially important silicon and lead compounds are described in separate sections below (see pp 136 and 131).

By far the largest-scale manufacture of organometallic compounds concerns Group IVB elements. Organosilicon compounds, many of which usefully combine high thermal stability with low reactivity, are made on a large scale for the manufacture of silicone polymers. Organotin compounds are used widely, though in much smaller quantities (about one tenth of the silicone production), for biocidal and polymer-stabilization purposes and as polymerization catalysts. For example, trialkyltin oxides, acetates or other derivatives R_3SnX have found extensive application as agricultural fungicides, maximum activity occurring when the *total* number of carbon atoms in the three alkyl groups is 9–12, regardless of whether the alkyl groups are the same or different. Trialkyls and dialkyls R_2SnX_2 are used in wood and paint preservatives, in moth proofing, to prevent slime in paper manufacture, and to kill tapeworm in domestic animals. Dialkyls are catalysts for the preparation of polyurethanes, and stabilizers for polyvinyl chloride or rubbers, since they function as anti-oxidants and as screens of ultraviolet radiation and are reactive towards such products of decomposition of these materials as hydrogen chloride, free radicals or unsaturated compounds which might otherwise propagate further decomposition. Organolead compounds have similar uses (and are generally toxic), although currently the outlet which makes them the biggest tonnage organometallic product of all is the use of tetra-ethyl-lead as an antiknock agent for petrol (U.S. ca. 300 000 tons per year).

Organo-derivatives of the Group VB elements by contrast find fewer uses. Organoarsenic compounds are of historic importance as the first class of compound to be studied systematically as chemotherapeutic agents—the first even moderately effective treatments for syphilis and sleeping sickness, for example, were organoarsenicals developed during P. Ehrlich's classic researches earlier this century—and, together with organoantimony compounds, they still find some biocidal uses, although they are progressively being superseded by other less generally toxic materials. (All the organo derivatives of arsenic and antimony should be regarded as toxic). Organobismuth compounds have found no applications which were not served better by arsenic or antimony compounds.

Group IVB elements

Stabilities and reactivities. All the elements of Group IVB form a large number of organic derivatives, particularly of the type R_nMX_{4-n}, where R = alkyl or aryl, M = Si, Ge, Sn or Pb, and X = a monovalent atom or group such as H, OR, NR_2, halogen, or alkali metal; n = 1, 2, 3 or 4. Many derivatives $R_{2n+2}M_n$ or $(R_2M)_n$ containing M—M links are also known. The fully substituted tetra-alkyls or -aryls R_4M differ from the fully alkylated (or arylated) derivatives R_3M' of the neighbouring groups in their relatively low chemical reactivity. Thus tetramethyltin is unreactive to air and moisture, in strong contrast to trimethylindium and trimethyl-stibine, both of which inflame in air. Moreover, the low reactivity of tetramethylsilane, which allows it for example to find wide application as an internal nuclear magnetic resonance standard (it absorbs at higher field than most organic substances), may be contrasted with the high reactivity of trimethylaluminium or trimethylphosphine. The reactivity of the Group III organic compounds may be attributed mainly to their vacant orbital and acceptor character. That shown by the trivalent Group V organic compounds is to a large extent due to their unsaturated character and to the presence of a lone-pair of electrons. The Group IVB alkyls and aryls behave as saturated compounds, and the elements show little tendency to expand their covalency above four unless they are bound to strongly electronegative atoms or groups; e.g. whereas tetramethyltin is unreactive, tin(IV) chloride is readily hydrolysed and also gives rise to compounds like ammonium chlorostannate $(NH_4)_2SnCl_6$. Electron-deficient bridging, whether by hydrogen or organic groups, is unknown in Groups IV and V as a means of association, although bridged species may well be involved as transient intermediates in exchange reactions.

Within Group IVB, the reactivity of the M—C bond in compounds R_4M increases progressively from carbon to lead as the bond energy decreases, and as expansion of the co-ordination number of M becomes easier with increasing size and decreasing difference between np and nd energy levels. Mean M—Me bond dissociation energies are C, 83; Si, 70; Ge, 59; Sn, 52; and Pb, 36 kcal/mole (see Table I, p 4). The thermal stabilities of R_4M decrease in the sequence $R_4Si > R_4Ge > R_4Sn > R_4Pb$. Relative reactivities are illustrated by the action of chlorine on the tetra-ethyls: Et_4C and Et_4Si undergo atomic chlorination with retention of the Et—C or Et—Si bonds, Et_4Ge is easily broken down to $Et_3GeCl + EtCl$, Et_4Sn reacts so fast that care is needed to stop at Et_3SnCl, and Et_4Pb is completely decomposed. Differences in M—C bond polarities (due to differences in the electronegativities of M) appear to have little influence on the relative reactivities of R_4M, but are difficult to assess in view of uncertainties about the relative positions of silicon, germanium, tin and lead on electronegativity scales, compare the sequences $C > Ge > Si \sim Sn > Pb$ (Allred-Rochow) and $C > Pb > Ge > Sn > Si$ (Pauling).

Lower oxidation states—catenation. An important feature of the organic chemistry of Group IVB elements is the large number of catenated derivatives, compounds containing metal-metal bonds. Though metal-metal bonds are of course present in all metals themselves, and are relatively common in the lower oxidation state chemistry of many transition metals, they are rare in compounds of main group metals outside Group IV with the important exceptions of the mercurous ion, As, Sb and Bi. Thermochemical studies on the Group IV derivatives have given conflicting values for M—M bond dissociation energies, but it is clear that these decrease down the group as do the thermal stabilities of derivatives R_6M_2. Ease of chemical cleavage of M—M bonds also increases down the group, and the range of derivatives known to contain such links increases in the sequence $Pb < Sn < Ge < Si < C$. Thus silicon forms catenated halides Si_nX_{2n+2}; ($X=Cl$, F), hydrides Si_nH_{2n+2} and organo derivatives Si_nR_{2n+2}; germanium forms fewer halides Ge_nX_{2n+2} but many hydrides Ge_nH_{2n+2} and alkyls Ge_nR_{2n+2}. The compounds Sn_2Cl_6 and Sn_2H_6 are the only catenated tin halides and hydrides to have been made but many linear or cyclic organo-polystannanes Sn_nR_{2n+2} or $(SnR_2)_n$ have been prepared, some with partial substitution of R by H or Cl. Lead-lead links are weak, and few compounds contain more than one of them—the neopentane analogue $(Ph_3Pb)_4Pb$ is the most complicated example to have been characterized to date (1967). The only other known examples have organic groups, not hydrogen or halogen, as substituents.

The effect of substituents on the thermal stability of catenated derivatives is worth noting: stability increases in the sequence halide < hydride < organo. Significantly, the substituents which stabilize catenated compounds are not those which form the strongest bonds to the Group IV elements, but those which release electrons most readily. It is likely that the M—M bond energy in a derivative $[X(\delta-)]_3M(\delta+)\cdot M(\delta+)[X(\delta-)]_3$ is sensitive to the size of the positive charge $\delta+$ on M, and if, as seems likely in many cases, fission of the M—M bond is a key step in their decomposition, the stability sequence is readily understood.

Apart from these catenated derivatives, many related compounds such as $R_3GeSnR'_3$ with two different Group IV metals directly linked are known, and an extensive chemistry of compounds having groups R_3M attached to many transition elements is currently being explored. Some of these derivatives, formal analogues of alkyl (R_3C) derivatives, possess useful catalytic properties in such reactions as polymerizations and hydrogenations. Compounds containing $Pt—SnCl_3$ structural units are particularly noteworthy in this respect.

In catenated derivatives, the Group IV elements retain a covalency of four although the formal oxidation number ranges from $+3$ for ethane-like compounds $R_3M\cdot MR_3$ to $+2$ for the cycloalkane analogues $(R_2M)_n$. Indeed, the organic chemistry of all the Group IVB elements is dominated by tetra-covalency; the only divalent monomeric organic derivatives of M(II) yet prepared are the cyclopentadienyls $(C_5H_5)_2Sn$ and $(C_5H_5)_2Pb$, which have sandwich structures 4.1 (electron diffraction), the angular shape of which arises from the stereochemically significant lone-pair.

C—C	$1\cdot431\pm0\cdot009$ A	C—C	$1\cdot430\pm0\cdot006$ A
C—H	$1\cdot142\pm0\cdot056$ A	C—H	$1\cdot105\pm0\cdot018$ A
Sn—C	$2\cdot706\pm0\cdot024$ A	Pb—C	$2\cdot778\pm0\cdot016$ A
α	ca. 125°	α	135°±15°

4.1

The lone-pair appears frequently to be structurally significant in the inorganic compounds of Ge(II), Sn(II), and Pb(II), oxidation states which are of course of much more importance in the inorganic chemistry of these elements, particularly tin and lead.

Co-ordination number. The group covalency of four allows a wide range of possible organo derivatives (R_4M, R_3MX, R_2MX_2, RMX_3, etc.) and this, coupled with ease of manipulation in comparison with most other types of organometallic compound, has led to the preparation of very large numbers of organo derivatives. Although four-covalency is a feature of virtually all the silicon and germanium compounds, and electron-deficient bridging is unknown for Si, Ge, Sn or Pb, association through functional groups resulting in co-ordination numbers greater than four is common in organotin and organolead chemistry. For example, trigonal bipyramidal five-co-ordination occurs in many trimethyltin derivatives Me_3SnX, in which planar (equatorial) Me_3Sn units are bridged by electronegative (axial) groups X which may be F, OH, O_2CR or even such species as BF_4 and ClO_4. Unco-ordinated cations R_3Sn^+ are unknown. The chain structure of the fluoride Me_3SnF, determined by an X-ray diffraction study, is shown in Figure 24.

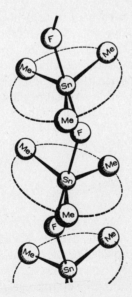

Figure 24. The structure of the trimethyltin fluoride polymer. (After H. C. Clark, R. J. O'Brien and J. Trotter, *J. Chem. Soc.*, 1964, 2332).

Trimethylgermyl fluoride, by contrast, is monomeric. Dimethyltin difluoride Me_2SnF_2 in the crystal contains octahedral tin, with a *trans* arrangement of methyl groups above and below each tin in the $(SnF_2)_n$ plane, itself composed of an infinite two dimensional network of tin atoms and bridging fluorines. This is like the crystal structure of tin tetrafluoride, but with the non-bridging fluorine atoms of SnF_4 replaced by methyl groups. The chlorides Me_3SnCl and Me_2SnCl_2 are not associated, but are volatile and readily hydrolysed, probably *via* five-coordinate intermediates, Me_nSnCl_{4-n},H_2O, to Me_3SnOH and $Me_2Sn(OH)_2$. The monochloride (unlike Me_4Sn) is a Lewis acid and forms adducts like the 5-co-ordinate pyridine complex 4.2

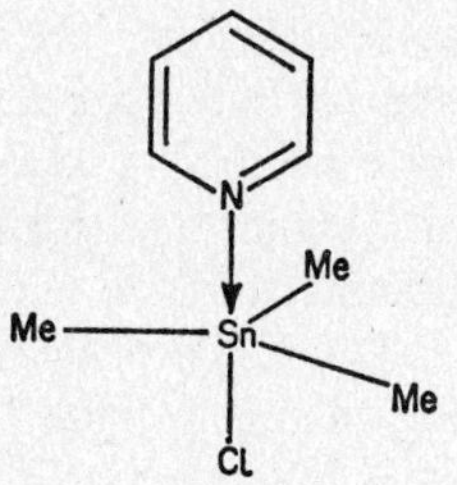

4.2

Lead probably has an even greater tendency than tin to expand its co-ordination number above four, and bridging occurs even in the chlorides. Thus diphenyl-lead dichloride Ph_2PbCl_2 has the octahedral arrangement of groups about each lead atom shown in 4.3, with a square planar arrangement of chlorine atoms, and phenyl groups occupying the axial positions. All the chlorine atoms are thus bridging atoms, linking the linear Ph_2Pb units into infinite chains.

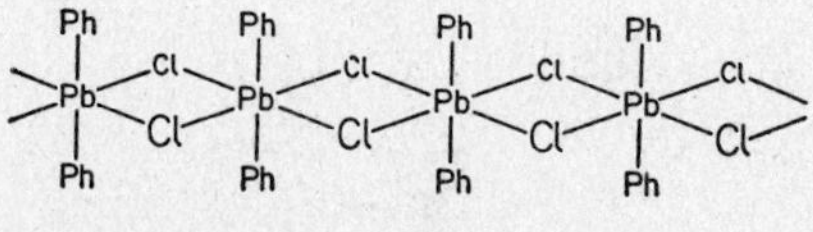

4.3

An illustration of the relative tendencies of germanium and tin to form five-co-ordinate species is provided by the crystal structures of the cyanides Me_3GeCN and Me_3SnCN. In the crystal, trimethylgermyl cyanide molecules have the expected tetrahedral geometry 4.4, but the alignment of the molecules is based on linear $\cdots Ge-C\equiv N \cdots Ge-C\equiv N \cdots$ chains (N $\cdots$ Ge distance 3·57 Å, cf. sum of N and Ge covalent radi

$3 \cdot 5$ Å) and with slightly flattened Me_3Ge units ($\angle Me$—Ge—Me $= 115°$). Trimethyltin cyanide by contrast is polymeric, with trigonal bipyramidal co-ordination about tin (4.5; Sn—C (or N) $= 2 \cdot 49$ Å, $\angle Me$—Sn—Me $= 120°$).

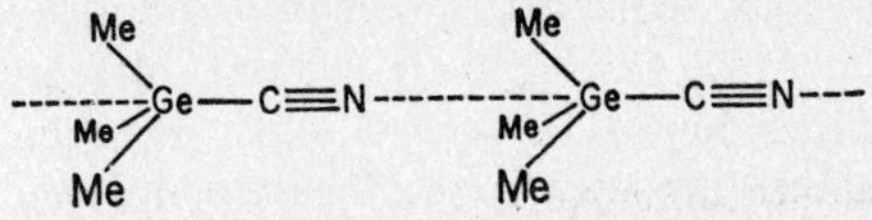

4.4 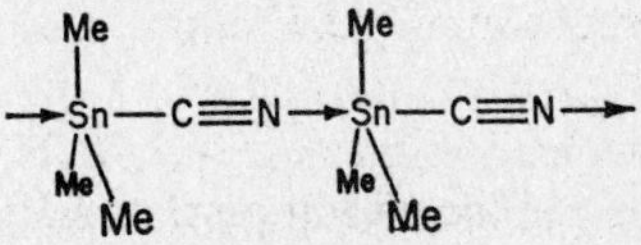   4.5

Five-co-ordinate silicon has recently been demonstrated in a few compounds, one the dimethylaminosilane pentamer $(H_3SiNMe_2)_5$, in which planar SiH_3 units are bridged by axial dimethylamino groups in a nearly planar ten-membered ring (4.6), another the cage compound 4.7, in which there is slight distortion of the trigonal bipyramidal arrangement of atoms about silicon because of the geometry of the triethanolamine residue (angle Ph—Si—O $= 97 \cdot 5°$). A further example is the pentafluorosilicate $[(Et_3P)_2PtCl(CO)]^+$ SiF_5^-. Hexafluorosilicates, e.g. K_2SiF_6, are of course relatively common.

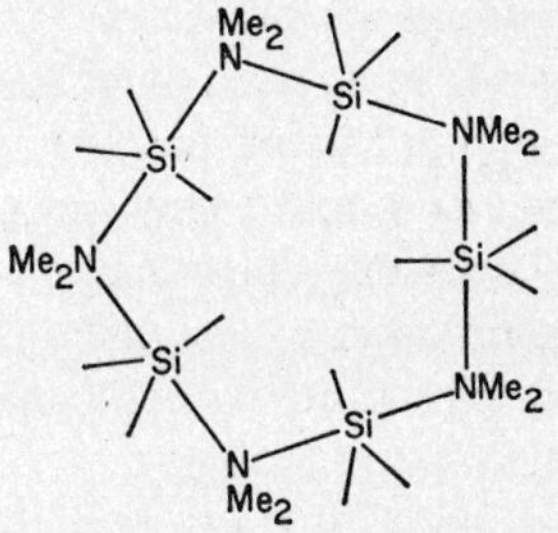

4.6 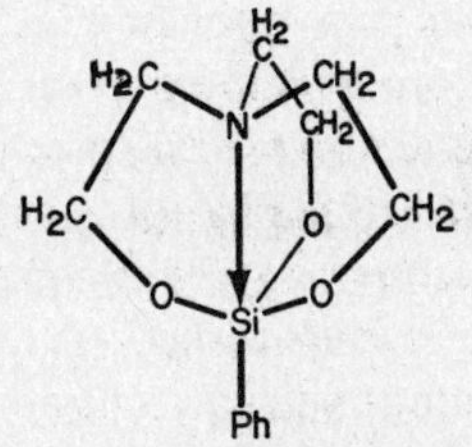   4.7

In these derivatives with co-ordination numbers greater than four, sp^3d or sp^3d^2 hybridization may be invoked to describe the bonding, although the extent of d orbital involvement is uncertain. The vacant d orbitals of Group IV elements in derivatives MX_4 thus appear to play a similar role to the vacant p orbital of boranes BX_3, in that they confer σ acceptor properties on MX_4, although in the case of the fully alkylated species R_4M the Lewis acidity is negligible. Another point of similarity is that they are available for dative π-bonding, $(p \to d)\pi$, between substituents such as nitrogen or oxygen and the Group IV element.

π-Bonding to Group IVB elements. The presence of multiple $(p \to d)$ π-bonding between electronegative atoms such as nitrogen or oxygen and silicon has been inferred in a variety of compounds by studies of electronic spectra, by structural studies (using infrared, Raman, n.m.r., microwave, and electron and X-ray diffraction techniques), and by studies of chemical reactions. The configuration of nitrogen attached to silicon can be particularly informative. Thus trisilylamine $(SiH_3)_3N$ is planar, like the triborylamine $(C_6H_4O_2B)_3N$, in which there is $N \rightleftharpoons B$ $(p \to p)\pi$ bonding, but unlike pyramidal Me_3N, in which the lone pair is localized on nitrogen. In the planar three-co-ordinate nitrogen, the electron pair not involved in σ-bonding to silicon may be regarded as occupying a p orbital which can overlap with vacant d orbitals on the silicons (4.8) to a greater extent than would be possible with an sp^3 hybrid nitrogen orbital. It should be stressed that multiple bonding is still *possible* in pyramidal molecules. Thus the pyramidal skeleton of $(GeH_3)_3P$ as opposed to planar $(SiH_3)_3P$ does not mean that $(p \to d)\pi$ interactions between germanium and phosphorus in trigermylphosphine are necessarily negligible.

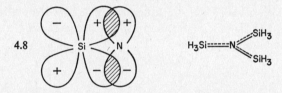

4.8

The linear skeletons of H_3SiNCO and H_3SiNCS, and near linear geometries of their methyl derivatives Me_3SiNCO and Me_3SiNCS, are also taken as indicative of $N \rightleftharpoons Si$ $(p \to d)\pi$-bonding. Similar $(p \to d)\pi$-bonding is also possible in compounds with silicon attached to oxygen, fluorine or carbon; for example, the bond angle at oxygen in $(Me_3Si)_2O$ is ca. 130°, and the electronic spectra of vinyl and acetyl derivatives $R_3SiCH:CH_2$ and $R_3SiCOCH_3$ are consistent with multiple $C \rightleftharpoons Si$ bonding.

A chemical method of detecting dative π-bonding to Group IV elements is to measure the relative donor properties of compounds in which such bonding is suspected, such as amines $(R_3M)_3N$ or oxides $(R_3M)_2O$. A useful guide to the Lewis basicity of such derivatives is provided by the shift they cause in $\nu(C-D)$ in the infrared spectrum of deuteriochloroform when they are mixed with $CDCl_3$, a shift due to $C-D---O$ or $C-D---N$ hydrogen bonding. Using this shift, trimethylgermyl compounds for example have been found to have donor properties comparable to their *t*-butyl analogues, greater than trimethylsilyl compounds but less than trimethyltin compounds (see Table X). The figures in Table X reflect

Table X. *Relative donor properties of some organometallic bases*
$\Delta\nu$ (cm^{-1}) for ν(C—D) of CDCl$_3$

Base	C	Si	Ge	Sn	Pb
Me$_3$MNEt$_2$	~100^a	64	82	90	—
(Me$_3$M)$_3$N	~100^a	0	72	106	—
Me$_3$MOEt	29	21	38	56	—
(Me$_3$M)$_2$O	33	13	55	84	—
Me$_3$MSMe	33	29	34	36	49
(Me$_3$M)$_2$S	40	29	38	43	51

^{a}Estimated.

differences both in the inductive effects (electronegativities) of the Group IV elements and also in their capacity to form $(p \rightarrow d)\pi$-bonds with oxygen, nitrogen or sulphur. Note the absence of interaction between (Me$_3$Si)$_3$N and CDCl$_3$, implying complete involvement of the nitrogen 'lone-pair' in π-bonding, in contrast to the strongly basic (Me$_3$Sn)$_3$N, which is pyramidal, and in which N $\rightleftharpoons$ Sn multiple bonding, if present at all, does little to offset the electron-releasing properties of the trimethyltin group.

Other chemical consequences of multiple bonding to silicon include the greater acid character of trimethylsilanol Me$_3$SiOH than of trimethylcarbinol Me$_3$COH (the opposite would be expected taking into account only inductive effects, but multiple bonding Me$_3$Si$\rightleftharpoons$OH reduces the negative charge on oxygen), and also the σ-electron releasing (+I) and π-electron withdrawing (–T or –M) effects of the trimethylsilyl group attached to aromatic systems, effects which have been confirmed by extensive studies of the acid or base strengths or directing effects of appropriate derivatives.

Despite the evident importance of $(p \rightarrow d)\pi$-bonding in organosilicon chemistry, and its occurrence to a limited extent in organogermanium chemistry, no compounds are known in which $(p—p)\pi$-bonding occurs significantly to Group IV elements other than carbon. This difference between the first and lower members of a group is typical, and is probably a consequence of the greater size and less effective overlap of the p orbitals of the heavier elements (see, e.g. the comparison of boron with aluminium on p 36). One most important consequence in organosilicon chemistry is that silicones (R$_2$SiO)$_n$ are polymers based on (—Si—O—)$_n$ chains, not monomers like ketones (see p 137). Another consequence is that organosilicon free radicals such as Ph$_3$Si·, for which C=Si multiple bonding might be a source of stability (as in 4.9 and 4.10), are in fact highly reactive; Ph$_3$Si· is much more reactive than Ph$_3$C· .

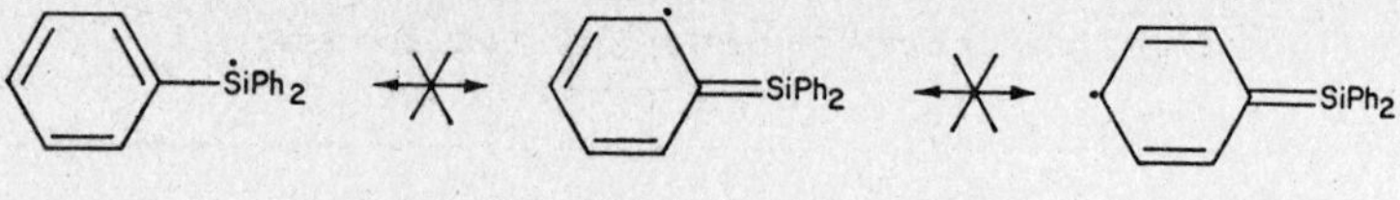

$$4.9 \qquad\qquad\qquad\qquad 4.10$$

However, an interesting feature of the thermal decomposition of organosilanes, which involves free radicals, is the number of products which apparently result from reactions involving diradicals. For example, the pyrolysis of tetramethylsilane at 700° for 1 minute leads to a complex mixture of carbosilanes, compounds with $-Si-CH_2-Si-$ skeletons, including a significant proportion ($\sim 7\%$) of what is effectively the parent compound $Me_3SiCH_2SiMe_3$, and surprisingly large proportions ($> 10\%$) of cyclic derivatives such as 4.11 and 4.12, which are believed to be formed *via* diradicals $Me_2\overset{\bullet}{Si}-\overset{\bullet}{CH_2}$. If these diradicals are indeed the intermediates,

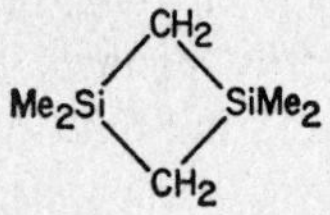

$$4.11 \qquad\qquad\qquad 4.12$$

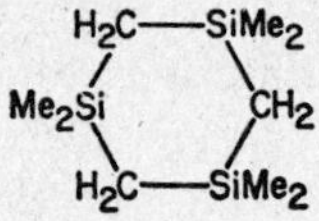

their apparent abundance is difficult to understand, unless they derive some stabilization as $Me_2Si{=}CH_2$.

Carbosilanes, with their $Si-CH_2-Si$ skeletons, are formally analogous to silicones, which have $Si-O-Si$ skeletons (p 135). However, whereas silicones have found wide application, the less useful bulk properties of carbosilane polymers and their less readily controlled methods of synthesis have prevented their finding comparable use. Apart from the pyrolytic route already described, they can be prepared directly from silicon and di- or poly-chloro alkanes, by thermal rearrangement of disilane derivatives ($Si-Si-C \rightarrow Si-C-Si$), and by dehalogenation of chloromethylsilanes, e.g.

$$ClSiMe_2CH_2Cl \xrightarrow{\;Mg\;} (Me_2SiCH_2)_2 \xrightarrow[\text{catalyst}]{\;Pt\;} (Me_2SiCH_2)_n$$

Preparative aspects. The most convenient methods for attaching organic groups to silicon, germanium, tin and lead in the laboratory employ Grignard, organolithium or organoaluminium reagents. Industrially, the action of an alkyl or aryl halide on the Group IV element alloyed with copper, sodium or magnesium is usually more economic. The industrial preparation of organosilicon compounds is described on p 136. The industrial preparation of tetramethyl- and tetraethyl-lead for use mainly as antiknock additives for petrol generally employs the reaction between methyl or ethyl chloride and a lead-sodium alloy (composition about NaPb):

$$4EtCl + 4NaPb \longrightarrow Et_4Pb + 3Pb + 4NaCl$$

After addition of water the product is separated by steam distillation and the metallic lead reconverted into NaPb and re-cycled. The need to recycle so much lead is an unattractive feature of the process, but sodium-rich alloys such as Na_4Pb do not react satisfactorily, and considerable effort has been devoted to processes free from this defect. Conversion into Et_4Pb of roughly half of the lead in the starting materials can be achieved in reactions between Et_3Al and PbO, PbS, $Pb(OAc)_2$ or $PbSO_4$, but the only other method used industrially involves the electrolysis of a solution of a Grignard reagent using a lead anode and an inert cathode:

$$4RMgX + Pb \longrightarrow R_4Pb + 2Mg + 2MgX_2$$

Organotin compounds are accessible by reactions between sodium-tin or magnesium-tin alloys and alkyl halides, by electrolytic methods using tin anodes, and by the action of organo-lithium, -magnesium, or -aluminium reagents on tin(IV) chloride. When the Grignard method is used, the vigour of the reaction can be abated by dissolving the halide in benzene or toluene. An excess of Grignard reagent is needed to minimize the formation of R_2SnCl_2 and R_3SnCl, which however may be removed by shaking the ether solution with aqueous alcoholic KF (which precipitates R_2SnF_2 and R_3SnF) or by adding dry ammonia, which precipitates adducts $R_3SnCl,2NH_3$, etc. When organo-aluminium reagents are used, complexes such as $R_2SnCl_2,AlCl_3$ or $R_3SnCl,AlCl_3$ are formed which hinder further alkylation of the tin. These complications can be avoided by the addition of donor species (NaCl, R_2O, R_3N) which complex with the aluminium halide as it is formed, leaving the alkyltin halides free to react further to the tetra-alkyl stage:

$$4R_3Al + 4NaCl + 3SnCl_4 \longrightarrow 3R_4Sn + 4NaAlCl_4$$

Organotin halides R_nSnX_{4-n}, which are valuable intermediates in the preparation of many organotin compounds, are accessible not only from RX and Sn or Na/Sn at 200–300°, but also by exchange reactions between R_4Sn and SnX_4. These exchange reactions are stepwise and so allow formation of mono-, di- or tri-alkyls:

$$R_4Sn + SnCl_4 \xrightarrow[0-20°]{\text{rapid}} R_3SnCl + RSnCl_3 \xrightarrow[180°]{\text{slow}} 2R_2SnCl_2$$

The wide range of derivatives which can be prepared from the halides are illustrated (for trialkyl derivatives) in the following diagram. Organogermanium and -silicon halides undergo similar reactions.

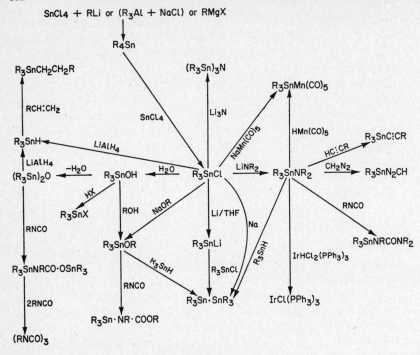

Some reactions

(a) Reactions involving M—C bond cleavage. The low polarity of the metal-carbon bond in organo-derivatives of Group IVB elements makes them less reactive and less generally useful for the alkylation of metal halides than such derivatives of more electropositive elements as RLi, RMgX and R$_3$Al. Accordingly, relatively few reactions are known in which organic groups are usefully transferred to another element from silicon or germanium. Organotin compounds find more uses, because the relative weakness of the Sn—C bond (ca. 50 kcal/mole) offsets its low polarity.

An example of the use of alkyl derivatives of silicon for the preparation of another organometallic compound is the preparation of methylgallium dichloride, (MeGaCl$_2$)$_2$ from gallium trichloride and (Me$_3$Si)$_2$O or Me$_4$Si. Tetramethylgermane and tetramethyltin react similarly, and progressively more readily:

$$Me_4M + GaCl_3 \longrightarrow MeGaCl_2 + Me_3MCl \; (M = Si, Ge \text{ or } Sn)$$

Organotin compounds are particularly useful for such purposes as the preparation of vinyl-lithium (from vinyl$_4$Sn + PhLi), of phenylboron chlorides (from 4BCl$_3$ + 1 or 2 Ph$_4$Sn), and of fluorocarbon derivatives, e.g.

$$Me_3SnCF_3 \xrightarrow{BF_3} Me_3SnCF_3, BF_3 \xrightarrow{KF} Me_3SnF + K^+[CF_3BF_3]^-$$

Cleavage of the M—C bond may occur more readily if the organic group has electronegative substituents, as these facilitate separation of the group as R^- during nucleophilic attack on the metal. The reactivities of α-, β- and γ-halogenated alkylsilanes exemplify this, although the β-substituted alkyls are unusually reactive. Thus, chlorinated methyl groups are quite readily removed from silicon by bases such as OH^- or $NR_2{}^-$, reactivity increasing in the sequence $CH_2Cl < CHCl_2 < CCl_3$. Cleavage of β-chloroalkyl groups occurs very readily, often for example when the distillation of compounds containing such groups is attempted, or by action of alkali:

$$Et_3SiCH_2CH_2Cl + OH^- \longrightarrow Et_3SiOH + Cl^- + CH_2{:}CH_2 \uparrow$$

γ-Chloro compounds form cyclopropane rings:

$$R_3SiCH_2CH_2CH_2Cl + OH^- \longrightarrow R_3SiOH + Cl^- + \overline{CH_2CH_2CH_2} \uparrow$$

(b) *Reactions of M—H, M—O and M—N bonds.* The relatively low reactivity of the M—C bond of Group IV alkyls and aryls allows organic groups to be used as 'blocking' groups to occupy spare valencies and so stabilize and simplify the chemistry of hydride, alkoxy or amino derivatives. Trialkyl or triaryl derivatives R_3MH, R_3MOR or R_3MNR_2 are normally much more convenient materials to work with than MH_4, $M(OR)_4$, or $M(NR_2)_4$, for reasons of thermal stability, solubility and reactivity. The thermal stabilities of the hydrides of a particular element, for example, increase in the sequence $MH_4 < RMH_3 < R_2MH_2 < R_3MH$, and whereas insertion of unsaturated substrates into the M—H, M—O or M—N bond of species R_3MH, R_3MOR or R_3MNR_2 can reasonably be expected to approach completeness of reaction, the tetra-hydrides, -alkoxides or -amides give complex mixtures of products.

The hydrides of silicon, germanium, tin and lead probably all have slight polarity $M(\delta+)$—$H(\delta-)$, and accordingly can add to polar unsaturated functional groups in a similar manner to boron hydrides. However, the polarity is particularly sensitive to the electron-releasing or -withdrawing properties of the other groups attached to M, and also to the polar nature of the solvent. Extensive studies on silicon and tin hydrides have shown these to be capable of adding to unsaturated substances by a variety of mechanisms, frequently by heterolytic fission of $M(\delta+)$—$H(\delta-)$, but also by homolytic fission into radicals and occasionally as if polarized $M(\delta-)$—$H(\delta+)$.

These hydrides are occasionally used to prepare hydrides of other elements, particularly those which might prove reactive towards, or difficult to separate from, polar reagents or solvents. Thus trimethylsilane and gallium trichloride give gallium dichloride hydride and trimethylchlorosilane:

$$Me_3SiH + GaCl_3 \longrightarrow 1/2(HGaCl_2)_2 + Me_3SiCl$$

The by-product Me_3SiCl is volatile enough to be distilled without difficulty from the reaction mixture. Reactions of various trimethylsilyl derivatives Me_3SiX, particularly where $X = H$, NR_2, OR, SR or pseudohalide, with covalent chlorides or bromides are frequently convenient methods of preparing such derivatives of other elements, e.g.

$$3Me_3SiNHPh + PBr_3 \longrightarrow 3Me_3SiBr + P(NHPh)_3$$

Organotin hydrides add to alkenes and alkynes by both polar and radical mechanisms, frequently concurrently, according to the conditions. Such additions when carried out between dienes and dihydrides R_2SnH_2 lead to polymers incorporating tin in the polymer backbone

$$CH_2{:}CH{\cdot}R{\cdot}CH{:}CH_2 + R'_2SnH_2 \longrightarrow (-R'_2SnCH_2CH_2{\cdot}R{\cdot}CH_2CH_2-)_n$$

The products of addition of tin hydrides to groups such as $C{=}O$ or $C{=}N$ have tin attached to the hetero atom, and are themselves susceptible to further reaction with tin hydrides, forming tin-tin bonds, e.g.

$$PhNCO + 2Ph_3SnH \longrightarrow PhNHCHO + (Ph_3Sn)_2$$

Materials $(R_2Sn)_n$ result from reactions of dihydrides R_2SnH_2. Suitable control of such reactions allows them to be used as routes to organo derivatives of higher stannanes, e.g.

$$2R_3SnN(Ph)CHO + R'_2SnH_2 \longrightarrow R_3SnSnR'_2SnR_3 + 2PhNHCHO$$

Condensations between dihydrides and diamides give cyclostannanes, $(R_2Sn)_n$ $(n = 6, 7, 8$ or $9)$:

$$Et_2SnH_2 + Et_2Sn(NEt_2)_2 \longrightarrow 2/n(Et_2Sn)_n + 2Et_2NH$$

This method can be extended to the formation of bonds between tin and transition elements by use of R_3SnNMe_2 and a metal hydride complex:

$$Me_3SnNMe_2 + \pi\text{-}C_5H_5W(CO)_3H \longrightarrow \pi\text{-}C_5H_5W(CO)_3SnMe_3 + Me_2NH$$

Other routes use the elimination of metal halide from Sn—Cl and Na—M or from Sn—Li/Na and Cl—M:

$$Ph_3SnCl + NaMn(CO)_5 \longrightarrow Ph_3SnMn(CO)_5 + NaCl$$

Organotin oxides, as well as alkoxy and amino derivatives, are capable of adding to various unsaturated substances including isocyanates, aldehydes and carbon dioxide. As the products may themselves be capable of further addition, the reaction can be applied to the oligomerization of the unsaturated compound, e.g.

Further examples of the versatility of amino derivatives R_3SnNMe_2 as reagents in organometallic syntheses include their use to prepare reactive cyclopentadienides or acetylides,

$$R_3SnNMe_2 + HC:CR' \longrightarrow R_3SnC:CR' + Me_2NH \uparrow$$

and their use as dehydrohalogenating reagents:

$$IrHCl_2(PPh_3)_3 + Me_3SnNMe_2 \longrightarrow IrCl(PPh_3)_3 + Me_3SnCl, NHMe_2$$

In this last capacity they frequently appear to function better than organic bases as they provide simultaneously an acceptor site (Sn) for the halide and a donor site (N) for the hydrogen. Amino derivatives of several other metals and metalloids react similarly. These and other reactions of organo-tin compounds are included in the diagram on p 132.

Silicones

The most important organosilicon compounds are the silicone polymers; in annual production (about 50 000 tons in the 'western world') they come second after organolead compounds, though the scale of organo-aluminium production, particularly in the U.S., seems to be catching up with that of silicones.

The most typical example of a silicone polymer is a 'methylsilicone oil'

$$Me_3Si \cdot (OSiMe_2 \cdot)_n OSiMe_3$$

in which n may lie between 20 and 500 according to the viscosity desired. Methylsilicone oils presently account for about one half of the total production of silicone polymers, the rest consisting of resins (for electrical applications) and rubbers, which are usable over a much greater temperature range than hydrocarbon rubbers. Silicone polymers had been made on many occasions before they became industrial products. For example Ladenburg made a silicone oil nearly a century ago (1872), and Kipping (in the period 1899–1944) made both oils and resins. Industrial development of these materials began, as commonly happens, only when the need arose to make a product with properties beyond those obtainable with currently available substances, and was much aided by concurrent clarification of the principles of polymerization (e.g. the work of Staudinger). In this particular instance the need was to develop an electrically insulating material which could be compounded with glass fibre and which would withstand substantially higher temperatures than the organic polymers known at that time (shortly before the 1939–45 war). Since organic polymers commonly produce carbon as a result of thermal breakdown, and as carbon conducts electricity the breakdown of an organic polymer used as electric insulator causes a rapidly deteriorating situation. Typical silicone polymers, however, are based on a silicon-oxygen chain or framework and tend to decompose thermally or by oxidation to silica and to volatile low molecular weight products which can readily escape from the site of the breakdown. The first large scale silicone product was a grease for coating aircraft sparking plugs, where the requirement was for high thermal stability and for absence of carbon formation in the event of flashover.

Silicone polymers were later found to have other properties (additional to high thermal stability) which now form the basis for most of the present production. These properties are, (1) unusually low temperature coefficients of viscosity, and (2) a range of surface effects.

Nearly all silicone polymers are made from the products obtained by hydrolysing organochlorosilanes, R_nSiCl_{4-n} in which R is most commonly methyl, and sometimes phenyl or hydrogen. Other groups may be introduced into the final polymer for special purposes. By far the greatest proportion of these chlorosilane intermediates consists of Me_2SiCl_2 (for the main element of most silicones), which is followed by $MeSiCl_3$ (for resins), $MeHSiCl_2$ (for establishing crosslinks and for attaching other molecules), Ph_2SiCl_2 and $PhSiCl_3$ (for resins) and Me_3SiCl (in relatively small amount for ending dimethylsiloxy polymer chains).

Whereas the foundations of organosilicon chemistry had been laid using the alkylation of $SiCl_4$ by zinc alkyls and later by Grignard reagents, the industrial development of silicone polymers stimulated the invention of cheaper routes.

The direct process for methylchlorosilanes. The 'direct process' in which methyl chloride reacts with a mixture of silicon and copper powder at 280–300° was discovered by E. G. Rochow (1945). The major reaction

$$2MeCl + Si \longrightarrow Me_2SiCl_2$$

is accompanied by the formation of small amounts of other and mostly useful products. A typical crude product composition, by wt%, is Me_2SiCl_2, 80; $MeSiCl_3$, 8; Me_3SiCl, 3; $MeHSiCl_2$, 3; $SiCl_4$, 1, various disilanes, 5. This process which gives a satisfactory distribution of products in much the proportions needed for the overall pattern of silicone manufacture, is still in use to-day with various modifications such as replacing mechanically stirred by fluidized bed reactors. The reaction was long believed to involve the transient formation of methylcopper, followed by attack of the resulting methyl radicals on subchlorides of silicon:

$$2Cu + MeCl \longrightarrow CuCl + MeCu$$
$$MeCu \longrightarrow Cu + Me\cdot$$
$$Si + CuCl \longrightarrow silicon\ subchlorides + Cu$$
$$silicon\ subchlorides + Me\cdot \longrightarrow methylchlorosilanes$$

A recent extensive study of the process (R. J. H. Voorhoeve, 1964) has shown that neither adding sources of methyl radicals nor adding radical scavengers has significant effect. The reactive solid phase is a copper-silicon alloy (about Cu_3Si in composition) in which the copper is positively

4.13   4.14

charged relative to the silicon. Thus methyl chloride would be adsorbed as in (4.13), and the rate-determining step is believed to be the nucleophilic displacement by chloride of the bonds between silicon and the underlying metal (4.14) leading to chemisorbed Me_2SiCl_2 and thence to free product. The importance of the relative charges on silicon and the other metal present is shown by the behaviour of silicides of calcium, chromium and iron in which the silicon is highly negative, about neutral, and positive respectively. Reaction of calcium silicide with methyl chloride gives predominantly Me_3SiCl since the large positive charge on the calcium results in its collecting more than one chlorine at a time, thus giving a preponderance of methyl on the silicon. Chromium silicide gives both Me_2SiCl_2 and $MeSiCl_3$ in comparable amounts, and iron silicide (on which the primary adsorption would have the chlorine on the silicon) gives mainly $MeSiCl_3$.

The hydrolysis of methylchlorosilanes. This goes rapidly and to completion, mainly since the Si—O bond energy (about 120 kcal/mole in siloxane polymers) considerably exceeds that of Si—Cl (about 85–90). The main points which determine the nature of the products are (a) the non-formation of Si=O double bonds, presumably due to poor $Si(3p)$—$O(2p)$ overlap and to the great stability of Si—O bonds, (b) the tendency of $\equiv$SiOH to eliminate water giving $\equiv$Si—O—Si$\equiv$. The bond energies of carbonyl bonds generally are in the range 160–175 kcal/mole, or twice (or more) that of C—O single bonds. The formation of $>$C=O and H_2O from $>$C(OH)$_2$ is thus normally favoured both on enthalpy and entropy grounds (formation of two molecules from one). Bond energies vary from compound to compound, but if one takes the lower end of the range (106–121) for Si—O, then that of Si=O would have to be over 210 kcal/mole for the formation of $>$Si=O from $>$Si(OH)$_2$ to be thermally neutral. Such a value approaches that of triple bonds between first row elements (N_2, 226) and is wildly improbable.

The reason for the condensation $2\equiv SiOH \rightarrow \equiv Si \cdot O \cdot Si \equiv + H_2O$, or $n\, R_2Si(OH)_2 \rightarrow (-R_2SiO-)_n + nH_2O$ is not so clear. Silanols can in fact be prepared and they vary greatly in the ease with which they eliminate water, a process which is catalysed by acids and by bases. The range of stability of silanols (generally mono > di > triols, and phenylsilanols > methylsilanols) is probably due to some extent to SiOH and SiOSi bond energy differences between one compound and another. Heats of combustion indicate that the mean SiO bond energy in a polydimethylsiloxane increases slightly with the chain length.

Hydrolysis of trimethylchlorosilane yields hexamethyldisiloxane,

$$Me_3SiCl + H_2O \longrightarrow H_3O^+ + Cl^- + Me_3SiOH$$

$$2Me_3SiOH \xrightarrow{\;H^+\;} Me_3SiOSiMe_3 + H_2O$$

the condensation of the intermediate trimethylsilanol being catalysed by the aqueous acid. The siloxane product is used to control the chain length of dimethylsiloxy polymers.

Hydrolysis of dimethyldichlorosilane is of course more complex, and gives a range of products. The first step

$$Me_2SiCl_2 + H_2O \longrightarrow Me_2ClSiOH$$

is followed by several further reactions, e.g.

$$2\,Me_2ClSiOH \xrightarrow{\;H_3O^+\;} Me_2ClSiOSiMe_2Cl$$

$$\downarrow H_3O^+$$

$$(Me_2ClSiOMe_2SiO-)_2 \xleftarrow{\;H_3O^+\;} HO\cdot Me_2SiOSiMe_2Cl$$

$$\downarrow H_3O^+ \qquad\qquad\qquad \downarrow H_3O^+$$

$$\begin{array}{c}\text{linear polymers}\\ HO(Me_2SiO)_nMe_2SiOH\end{array} \xleftarrow{\;H_3O^+\;} HO\cdot Me_2SiOSiMe_2OH$$

$$\downarrow -H_2O$$

$$\begin{array}{c} ^{a}Me_2Si\cdot O\cdot SiMe_2 \\ \;\;\overset{..}{O}\quad\;\;\overset{..}{O} \\ Me_2Si\cdot O\cdot SiMe_2 \end{array}$$

[a]The ring is puckered, but the SiOSi angles are quite large, 142°.

Normally a mixture of cyclic (exemplified above by 'cyclic tetramer') and linear oligomers is obtained containing between three and nine dimethylsiloxane (Me_2SiO) groups in each unit.

Since methanol can readily be converted to methyl chloride, the chlorine involved in the silicone process can be made to circulate, and the process then becomes in effect:

$$2MeOH + Si \longrightarrow \frac{1}{n}[Me_2SiO]_n + H_2O$$

Silicone oils. These are formed when the crude hydrolysis product mentioned above is made to undergo further polymerization using acid or alkali as catalyst. In this 'equilibration' process the hydroxy-ended oligomers condense with elimination of water, and the cyclic oligomers open and also produce long chain polymers. The average chain length is determined by the addition of exactly controlled amounts of hexamethyl-disiloxane:

$$x(-Me_2SiO-)_n + Me_3SiOSiMe_3 \longrightarrow Me_3SiO(-Me_2SiO-)_{nx}SiMe_3$$

The polymers thus obtained have not all the same chain length, there being a Gaussian distribution about the mean. As the polymers with relatively short chain lengths (less than Si_8) are undesirable for many applications (too volatile) they are removed by distillation after separation of the equilibration catalyst.

Probably the best known characteristic of methylsilicone oils is the low temperature coefficient of viscosity, this being normally one third to one quarter of that of a hydrocarbon oil of similar viscosity. The low temperature coefficient of silicone oils is believed to be due to two factors. The first of these is the free rotation of methyl groups about silicon coupled with the high flexibility of the siloxane chain leading to rather larger than normal intermolecular distances and smaller than normal intermolecular forces. A low temperature coefficient of viscosity implies low intermolecular forces, and large intermolecular distances agree also with the observed high degree of compressibility of silicone oils, and with their relatively high volatilities (i.e. low Trouton constants). The other effect believed to operate is a tendency for the polymer to form helical coils in which the electrostatic interaction between the polar SiO bonds is largely intra- rather than inter-molecular. Increasing temperature would cause gradual unwinding of the helices and an increase in the intermolecular forces partly offsetting the normal effect of temperature on viscosity. This somewhat resembles the use of polyisobutene and some other polymers to improve the viscosity-temperature characteristics of lubricating oils: the polymer tends towards a spherical shape at lower temperatures and unwinds as the temperature increases. Incidentally the 'silmethylene' or carbosilane polymers $(-Me_2SiCH_2-)_n$ have viscosity characteristics like those of hydrocarbon oils.

Though the viscosity properties are better known, the surface effects of silicones are considerably more important. About half of the total silicone production is in the form of oils which are used on account of a variety of surface effects. As silicones are relatively costly (roughly £1 per lb) it is fortunate that a lot of surface effect can be obtained for a given weight of polymer. All these effects are basically due to the presence both of highly polar SiO bonds and non-polar hydrocarbon groups in the same polymer. When textiles are treated with silicone oils the polymer chains appear to orient themselves with the polar groups close to the fibres and to present a hydrocarbon and hence water-repellent exterior. Silicone-treated textiles are not only water-repellent but also resist water-borne stains. The presence of some MeSiH groups (instead of Me_2Si) in the oil is advantageous because it allows the formation of cross-links which anchor the polymer to the fibres and give a wash-resistant finish. Closely related are the very extensive applications resulting from the anti-stick properties conferred on surfaces by suitable silicone treatment. Treated paper can easily be peeled away from, for example, an adhesive-coated wall covering. Nearly all vehicle tyres are made in moulds which have been silicone-treated to give easy release.

A major and at first sight perhaps a surprising application of silicone oil is as a lubricant for wax. The methyl oils do not dissolve in wax but become adsorbed on the surface of wax particles. Aqueous suspensions of wax containing 4–10% silicone can be rubbed out, e.g. on floor, furniture, or motor car, with much less effort than would be needed without the silicone lubricant.

Silicone rubbers. Rubbers in general are polymers with long chain lengths and in which there are occasional cross-links. This results in a high degree of deformation under stress, but without any flow. The introduction of increasing proportions of cross-links beyond the optimum (which varies from one system to another) results in diminished rubber-like properties and increased rigidity leading ultimately to hard resinous materials (discussed later).

The prime requirement for making silicone rubber is for a linear polymer $(R_2SiO)_n$ of longer chain length than obtains for any of the silicone oil products. This can only be made by very rigorous control of the proportion of chain-ending groups such as Me_3SiO. Normally Me_2SiCl_2 of a particularly high grade of purity is hydrolysed, or the cyclic tetramer $(Me_2SiO)_4$ is separated and carefully purified. This is in either case followed by a KOH equilibration step including the addition of a minutely controlled amount of $Me_3SiOSiMe_3$. The result is a Me_3Si-ended polymer about 6000–7000 siloxy units long. Shorter chains give rubbers that are too soft and sticky.

This high polymer gum is mixed both with a cross-linking agent and a filler such as very finely divided silica or carbon, and the cross-links are established by heating after the mixture has been formed in the desired shape by processes like rolling, extrusion or moulding. Though cross-links are commonly produced by the use of radical sources such as peroxides, other methods are sometimes used including the replacement of a few of the methyl groups in the siloxane gum by the more reactive vinyl group.

Silicone rubbers have various properties which cannot be matched by other rubbers, and which justify their relatively high price. Their flexibility at low temperature (even to $-80°$) and their useful life at $+250°$ of about 50–100 times that of hydrocarbon rubbers make them very suitable for numerous engineering applications, particularly in the aircraft industry. Their chemical inertness has resulted in many biological uses.

Silicone resins. These rigid polymers, produced by the hydrolysis of R_2SiCl_2 mixed with a substantial amount of $RSiCl_3$ and therefore extensively cross-linked, are used mainly for making electrically insulating material commonly as a glass-cloth laminate. They also have substantial uses on account of surface effects, namely as water-repellent treatment for masonry and for conferring relatively long-life easy-release properties on baking equipment used in the food industry.

Too high a proportion of $RSiCl_3$ in the R_2SiCl_2—$RSiCl_3$ mixture gives brittle or friable hydrolysis products of no practical use. Useful materials are obtained when the R:Si ratio is in the range $1·2–1·6:1$, the lower end of the range giving more highly cross-linked hard and glassy products, and the upper end giving more flexible and rubbery materials. Resins are among the small number of silicones in which it is advantageous to have some hydrocarbon groups additional to methyl bound to silicon, since co-polymers made from mixtures of methyl and phenylchlorosilanes have better properties in many respects than those derived only from Me_2SiCl_2 and $MeSiCl_3$.

Phenylchlorosilanes can be prepared by a direct process from silicon and chlorobenzene with copper or silver catalyst, but yields are not so good as those of methylchlorosilanes. Alternative routes are Grignard or phenylsodium arylation of $SiCl_4$, and the reaction between benzene and either H_2SiCl_2 or $HSiCl_3$ in the presence of platinum or peroxide catalysts. The required intermediates are made from silicon-copper (or-iron) and HCl:

$$Si + HCl \longrightarrow H_2SiCl_2 + HSiCl_3 + SiCl_4$$
$$HSiCl_3 + C_6H_6 \longrightarrow PhSiCl_3 + H_2$$

Resins are normally made in two or more stages. The methyl- and phenyl-chlorosilane mixture is first hydrolysed with a mixture of water and toluene. The product is a relatively low molecular weight oligomer with few cross-links. It is soluble in toluene, and at this stage can be washed free from chloride. After evaporation of much of the toluene the now viscous polymer can be used typically for the impregnation of glass-cloth and the polymerization completed by heating, finally under pressure at 250°.

Miscellaneous silicone products. Most of the major silicone manufacturers sell over 100 different silicone products, and the applications of these cover a very wide range. Here there is space to mention only two more applications: the first being the use of silicone oils and emulsions for the prevention of foam in numerous industrial processes, the other being the use of silicone-containing copolymers as surfactants in the manufacture of polyurethane foams from di-isocyanates, ethylene/propylene polyethers and a little water. Unless a suitable surface-active 'surfactant' is added, polyurethane foams tend to collapse before the polymerization is complete. One of the best types of surfactant is a polysiloxane to which polyethers are attached:

$$Me_3SiOSiMe(OSiMe_2)_xOSiMe(OSiMe_2)_yOSiMe(OSiMe_2)_zOSiMe_3$$
$$\overset{|}{R} \qquad\qquad \overset{|}{R} \qquad\qquad \overset{|}{R}$$

R = ethylene/propylene oxide polymer

Copolymerization of Me_2SiCl_2 and $MeHSiCl_2$ (one of the products from the direct process), using $Me_3SiOSiMe_3$ to control chain length, gives a polymer in which there are SiH in place of SiMe groups here and there along the chain. These SiH groups lend themselves very well to the attachment of side chains, typically by the reactions:

$$
\begin{array}{ll}
 & \xrightarrow{\text{HOCR}_3} Me\overset{|}{Si}OCR_3 \\
\overset{|}{\underset{|}{MeSiH}} & \\
 & \xrightarrow{\text{CH}_2=\text{CH}\cdot\text{R}} Me\overset{|}{Si}CH_2CH_2R
\end{array}
$$

Both reactions need catalysts such as tin(II) or platinum compounds. For some purposes one siloxane chain can be attached to another using the SiH/SiOH condensation:

$$
\overset{\vdots}{\underset{\vdots}{O}} \quad Me \qquad\qquad \overset{\vdots}{\underset{\vdots}{O}} \; Me
$$
$$MeSiH + HOSiO \cdots \longrightarrow MeSiOSiO \cdots$$
$$O \qquad Me \qquad\qquad O \; Me$$

In the manufacture of surfactants, the necessary hydroxy or vinyl group is attached to the end of a polyether chain. The vinyl method, which establishes a new SiC bond, gives a more hydrolytically stable product, since SiOC bonds are to varying degrees subject to hydrolysis.

Finally, though it has so far found little industrial use, 'bouncing putty' is worth a mention if only on account of its entertainment value. It is made by heating to ca. 200° a methyl silicone oil with about 5% of its weight of boric oxide and an inert filler. It slowly flows under its own weight, like a liquid, but bounces very well when formed into a ball and thrown at a hard surface. A very sharp blow can fracture a ball of bouncing putty.

Group VB elements

Types of compound. Arsenic, antimony and bismuth form organic derivatives in which their co-ordination number is three (e.g. R_3As), four (e.g. $R_4Sb^+X^-$), five (e.g. R_5Bi) and six in the particular case of antimony ($LiSbPh_6$). The co-ordination numbers of four and five are less common in the case of bismuth than of antimony or arsenic.

Apart from a wide range of tri-alkyls and -aryls R_3M, which have pyramidal structures, arsenic and antimony form organohalides in oxidation states $+5$ (R_nMX_{5-n}, where $n = 1$, 2, 3 or 4) and $+3$ (RMX_2 and R_2MX), and various oxygen derivatives including the arsonic acids $RAsO(OH)_2$, the weaker arsinic $R_2AsO.OH$, stibonic $RSbO(OH)_2$, and stibinic $R_2SbO.OH$ acids, and the oxides R_3MO, $(R_2M)_2O$ and $(RMO)_n$. Bismuth forms similar halides but, among its oxygen derivatives, no analogous acids.

The only organohydrides of arsenic, antimony and bismuth are derivatives of M(III), RMH_2 and R_2MH, and are volatile, thermally rather unstable substances, the antimony and bismuth compounds generally decomposing at or below room temperature. Several organoalkali metal derivatives of arsenic are known, including such types as R_2AsM, $RAsM_2$, $RAs(M)As(M)R$ and $RHAsM$. A few antimony and bismuth analogues of the first type have been prepared, including Bu_2SbLi, $(Bu_2Sb)_2Mg$ and Ph_2BiNa.

Donor-acceptor character and multiple bonding. Like phosphorus, arsenic shows pronounced donor-acceptor character in compounds of the type R_3As, and this is most marked towards transition elements like platinum capable of forming double $(d \rightarrow d)\pi$-bonds. Chelating and easily polarized diarsines such as *o*-phenylenebis(dimethylarsine), *o*-$C_6H_4(AsMe_2)_2$, form particularly stable complexes with many transition elements. The example quoted ('diars') has been widely used as a ligand, and is readily prepared from cacodylic acid, $Me_2AsO(OH)$:

$$Me_2AsO(OH) \xrightarrow{Zn/HCl} Me_2AsH \xrightarrow{Na/THF} NaAsMe_2 \xrightarrow{o\text{-}C_6H_4Cl_2} o\text{-}C_6H_4(AsMe_2)_2$$

Stibines R_3Sb have similar though weaker co-ordinating properties, but bismuth derivatives R_3Bi are very feeble complexing agents. This ability of arsines and stibines to form π-bonds to suitable acceptors is comparable to the capacity of the Group IVB elements to form π-bonds to such elements as nitrogen and oxygen (p 128), and is also manifest in the existence of the oxides $R_3M{:}O$, tosylimines $R_3M{:}NSO_2C_6H_4Me$, or Wittig reagents $R_3M{:}CR'_2$ as monomeric species in which oxygen, nitrogen or carbon is multiply-bonded to the Group V element. Multiple $As{\doteqdot}O$ bonding is also indicated by the As—O—As bond angle (137°

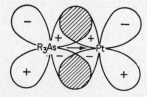 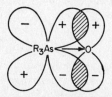

in the oxide $(Ph_2As)_2O$. It should be stressed that in all these cases the main group element is considered to use a suitable vacant d orbital. Like the Group IV elements, arsenic, antimony and bismuth show little tendency to form multiple bonds by use of their p orbitals. Thus oxides RAsO are polymeric, with As—O—As bridges, *not* monomers R—As=O, although because much of the organic chemistry of arsenic was developed before the significance of the state of association of such compounds was appreciated, they are still occasionally referred to as if they were monomeric. Another class of compound often referred to as if $(p{-}p)\pi$ bonding were present are arsenobenzenes $(RAs)_n$, which are frequently written RAs=AsR, but which are actually higher oligomers with As—As single bonds.

When three-co-ordinate arsenic atoms are incorporated in unsaturated ring systems, as in the five-membered arsole ring of 4.15 (preparable from $PhAsCl_2 + LiPhC{:}CPh{\cdot}CPh{:}CPhLi$), $(p \to p)\pi$ bonding between arsenic (if trigonally co-ordinated) and carbon would allow delocalization of the arsenic 'lone-pair' into the π-electron system of the ring, which would thereby acquire aromatic character (cf. cyclopentadienide $C_5H_5{}^-$ or pyrrole C_4H_4NH derivatives). Arsoles, however, are not aromatic. The crystal structure of 9-phenyl-9-arsafluorene (4.16) for example shows the arsenic to be pyramidally co-ordinated with the substituent phenyl group located in a plane approximately normal to the plane of the rest of the molecule. The stability of the arsenic pyramid is such as to allow resolution into optical isomers of unsymmetrically substituted arsafluorenes, such as the 2-amino derivative 4.17.

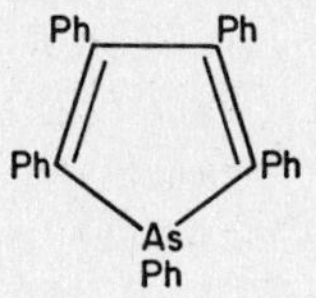

4.15 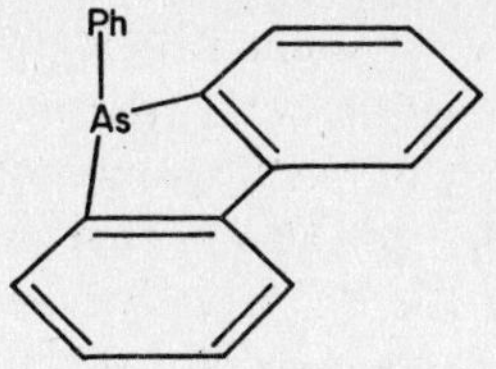   4.16 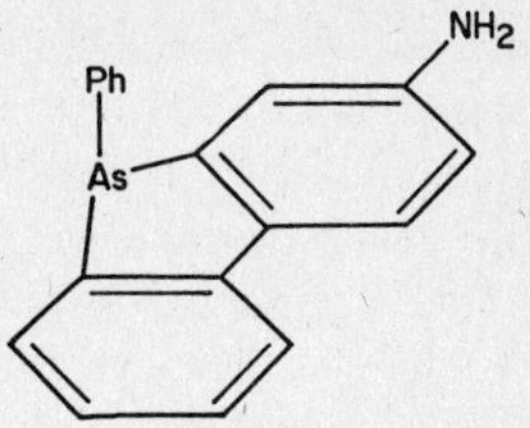   4.17

Arsenic, antimony and bismuth can all act as acceptors when bound to halogens. Again, use is presumably made of a vacant d orbital. An interesting example of the weak acceptor character of an arsine with only one electronegative substituent is provided by crystalline cyanodimethylarsine, Me_2AsCN, in which the pyramidal molecules are aligned with the nitrogen of one molecule near to the arsenic of its neighbour (As----N intermolecular distance 3·18 Å, cf. sum of van der Waals radii 3·5 Å, Compare also the structures of Me_3GeCN and Me_3SnCN, p 127).

Catenation. Metal-metal bonds are a feature of many organoarsenic compounds. For example, the first to be prepared (by Cadet de Gassicourt in 1760) was the spontaneously air-inflammable liquid 'cacodyl' which was later shown to be tetramethyldiarsine, $Me_2As·AsMe_2$, and several of the compounds found by Ehrlich to be useful chemotherapeutic agents were derivatives of arsenobenzene, $(PhAs)_6$. The structure of the parent compound has recently been shown to be based on a chair-shaped As_6 ring (4.18). Arsenomethane $(MeAs)_5$ is a pentamer with a puckered As_5 ring (4.19).

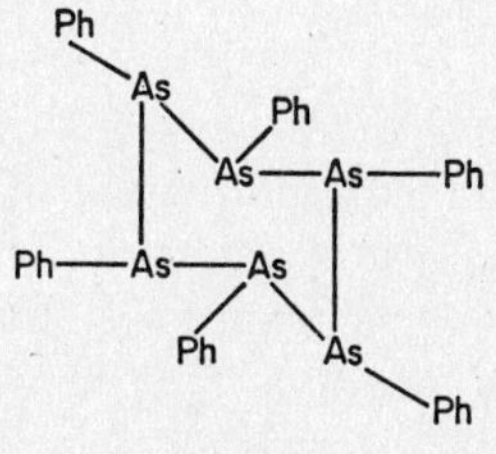

4.18 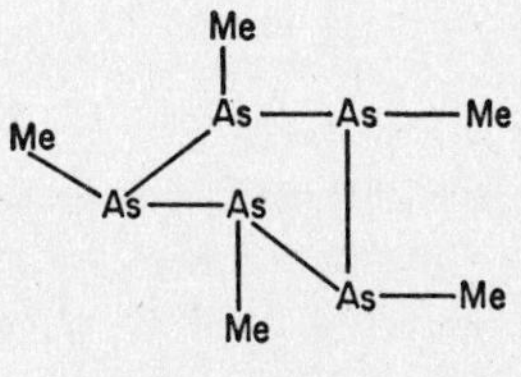   4.19

Metal-metal bond strengths decrease down Group VB, so Sb—Sb bonds are weak and Bi—Bi bonds even weaker. The known distibine derivatives $R_2Sb·SbR_2$ have as bismuth analogues only the very unstable tetramethyldibismuth and possibly a tetraphenyl derivative.

Onium salts. Arsenic, antimony and bismuth all form salts $R_4M^+X^-$ analogous to ammonium or phosphonium salts, and known respectively as arsonium, stibonium and bismuthonium salts. Their thermal stabilities and ease of preparation decrease markedly from arsenic to bismuth. Thus although trimethylbismuth does not add methyl iodide, trimethylstibine combines in the cold, forming $Me_4Sb^+I^-$, while Me_3As and Me_3P react progressively more quickly. Tetramethylarsonium iodide can even be prepared directly from arsenic and methyl iodide. Triaryl-arsines and -stibines do not take up aryl iodides, but the tetra-aryl salts $Ar_4M^+X^-$ can be made by the action of Grignard reagents on the oxides Ar_3AsO or Ar_3SbO or halides Ar_3MX_2. Mixed derivatives such as $Ph_3AsMe^+Br^-$ are similarly prepared. This last with phenyl-lithium gives triphenylarsine-methylene:

$$Ph_3AsCH_3{}^+Br^- + PhLi \longrightarrow Ph_3As{:}CH_2 + LiBr + PhH$$

Arsonium and stibonium salts have properties consistent with their ionic constitution. The hydroxides are strong alkalis like quaternary ammonium hydroxides. The tetrahedral structures of their cations have been confirmed in some instances by X-ray analysis. Of the few bismuthonium salts known, the tetraphenylborate $Ph_4Bi^+BPh_4{}^-$ (from $Ph_5Bi + Ph_3B$) is among the more stable. The halides readily lose halobenzene, e.g.

$$Ph_4Bi^+Cl^- \longrightarrow Ph_3Bi + PhCl$$

In contrast to these monohalides $R_4M^+X^-$, the dihalides R_3AsX_2 and R_3SbX_2 (from $R_3M + X_2$) are covalent derivatives with five-co-ordinate arsenic and antimony and a planar arrangement of the groups R about As or Sb. Five-co-ordination persists in aqueous solutions of these halides, which on the basis of their Raman spectra are believed to contain cations $[R_3M(OH_2)_2]^{2+}$ or $[R_3M(OH)(OH_2)]^+$, *not* R_3MOH^+ or $[R_3MOMR_3]^{2+}$ (cf. $[R_3Sn(OH_2)_2]^+$).

Penta-alkyls and aryls. The penta-alkyls and -aryls R_5M are of interest both structurally and in connection with the relative stabilities of the $+5$ oxidation state. Only antimony appears to form a pentamethyl derivative, Me_5Sb (a liquid bp 126–127°, from $MeLi + Me_3Sb$ or Me_4SbBr), which is surprisingly thermally stable, and unlike trimethylstibine it does not inflame in the air, though it oxidizes quickly and is decomposed by water. Its vibrational spectra are consistent with a trigonal bipyramidal structure. With an excess of methyl-lithium it forms the salt $Li^+SbMe_6{}^-$.

All three Group VB metals, like phosphorus, form crystalline penta-phenyl derivatives Ph_5M. An X-ray diffraction study of the antimony compound has shown a square pyramidal arrangement of phenyl groups about antimony as indicated in 4.20. The crystal structure of pentaphenyl-phosphorus, by contrast, has the expected trigonal bipyramidal arrangement of phenyl groups about phosphorus (4.21), and pentaphenylarsenic

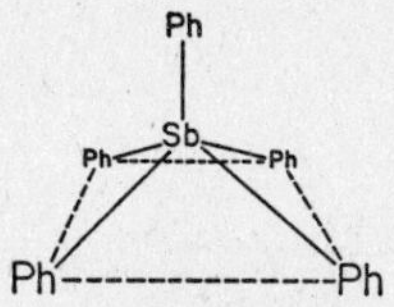

4.20 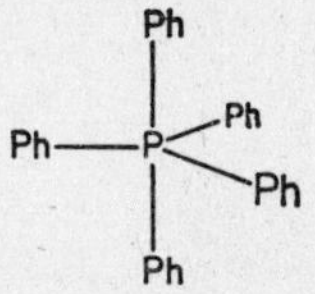   4.21

is isomorphous. However the structure of the thermally unstable bismuth analogue is not known. Pentaphenylantimony is a rare example ($InCl_5^{2-}$ is another) of a main group '10-electron' species with a square pyramidal structure, and this may well be a consequence of crystal packing considerations for this particular compound rather than an alteration in the relative stabilities of the two possible five-co-ordinate frameworks, which nevertheless are believed to differ in energy only slightly. The smallness of this energy difference has been demonstrated by ^1H n.m.r. studies on several penta-organo derivatives R_5M (M = P, As or Sb). Except in a few cases involving particularly bulky aryl groups, where rearrangements are sterically hindered, the substituents R have been found to appear to be equivalent, even at $-100°$, indicating rapid rearrangement from one trigonal bipyramidal (or square-based pyramidal) form to another. As axial-equatorial position changes of substituents in trigonal bipyramidal structures are generally believed to occur *via* square-based pyramidal intermediates, ready exchange can be taken as evidence of little energy difference between the two structures.

Preparative aspects. Since bonds between carbon and arsenic, antimony and bismuth are stable to water they may be made in aqueous solution. Thus, alkyl arsonic acids $RAs(OH)_2$ may be prepared by the Meyer reaction between sodium arsenite and alkyl halides:

$$RX + As_2O_3 \xrightarrow{OH^-} RAsO(OH)_2 + X^-$$

Their aryl counterparts are conveniently prepared by the Bart reaction in which diazonium salts are allowed to decompose in the presence of sodium arsenite:

$$PhN_2^+Cl^- + Na_3AsO_3 \longrightarrow PhAsO(ONa)_2 + NaCl + N_2$$

This reaction is also useful for attaching further aryl groups to arsenic,

K

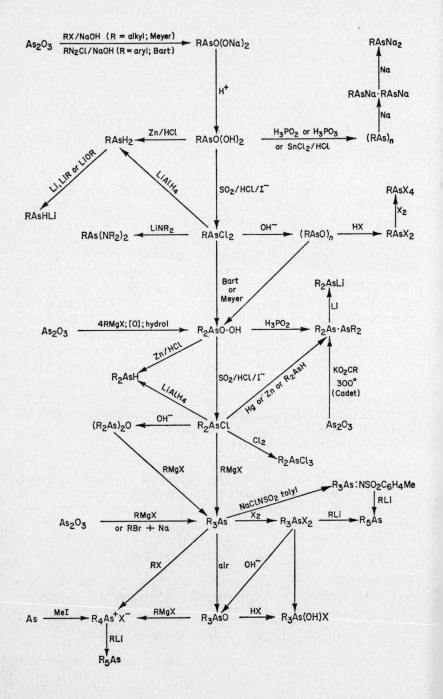

and for preparing aryl-antimony and -bismuth compounds. The original synthesis of an organo-arsenical was the preparation of cacodyl ($Me_2As \cdot AsMe_2$) and its oxide ($(Me_2As)_2O$) from arsenic (III) oxide and potassium acetate at 300°.

A few preparative routes to compounds with from one to five organic groups attached to arsenic and some of their reactions are illustrated in the diagram on p 148. The dominant positions of arsonic, $RAsO(OH)_2$, and arsinic, $R_2AsO(OH)$, acids and arsine oxides, R_3AsO, in these sequences reflect the ease of oxidation (e.g. by air) of As(III) to As(V). Arylarsonic acids are stable enough to be nitrated without cleavage of the As—C bond. Both series of acids are weak and the effect of successive replacement of OH groups of H_3AsO_4 by Me is illustrated by the acid dissociation constants: $AsO(OH)_3$, $K_1 = 5 \times 10^{-3}$; $MeAsO(OH)_2$, $K_1 = 2 \cdot 5 \times 10^{-3}$; $Me_2AsO(OH)$, $K = 7 \cdot 5 \times 10^{-7}$.

This last compound, cacodylic acid, is probably the best known arsinic acid, and is unattacked by fuming nitric acid, aqua regia or potassium permanganate, even on warming.

BIBLIOGRAPHY

Organosilicon compounds
C. Eaborn, 'Organosilicon Compounds' (Butterworth, London, 1960); E. A. V. Ebsworth, 'Volatile Silicon Compounds' (Pergamon, London, 1963).

Organohalosilanes and the reaction between methyl chloride and copper-silicon
R. J. H. Voorhoeve, 'Organohalosilanes—Precursors to Silicones' (Elsevier 1967).

Organogermanium compounds
F. Glockling, *Quart. Rev.* **20**, 1966, 45. The chemistry of germanium, tin and lead is presented in much greater detail in G. E. Coates, M. L. H. Green and K. Wade, 'Organometallic Compounds', 3rd Edition, Volume I, Chapter 4. (Methuen, London, 1967).

Catenation in Group IVB
H. Gilman, W. H. Atwell and F. K. Cartledge, *Advances in Organometallic Chemistry*, 1966, **4**, 1–94.

Organoarsenic compounds
W. R. Cullen, *Advances in Organometallic Chemistry*, 1966, **4**, 145–242.

CHAPTER FIVE

Organometallic compounds of the *d*-block transition elements: classification of ligands and theories of bonding

Classification of ligands

While the organometallic compounds of the main group metals are best classified by the Periodic Group of the metal, those of the *d*-block transition metals are more conveniently treated in terms of the organic ligands. In Table XI ligands are classified according to the number of electrons which are formally considered* to be donated by the ligand in the formation of the metal-ligand bond.

The 18-electron rule

As mentioned in Chapter 1, many transition metal complexes containing organic, carbonyl or hydride ligands, which are isolable at room temperature, can be regarded as having 18 electrons in the valence shell of the metal. This is the empirical basis of the rule—a valence shell containing

---

*The expression 'formally considered' is used not because there is any doubt with most ligands about the number of valence electrons involved in a given metal-ligand bond, but because there may be room for discussion about the polarity of the bond. For example the metal-hydrogen bond, M—H, may be thought of as being formed from $M^- + H^+$, $M^+ + H^-$ or $M\cdot + H\cdot$. Since the same bond is formed in all cases it makes no difference in the final counting of electrons what the preparative route is, or indeed whether the final polarity of the mainly covalent bond is $\overset{\delta+}{M}—\overset{\delta-}{H}$ or $\overset{\delta-}{M}—\overset{\delta+}{H}$. It is convenient to consider all M—H bonds in different covalent hydrides to be formally the same, and we choose the convention that the bonds arise from the radical H· and the metal atom, which, *with respect to this ligand*, has a neutral charge. Similarly in all alkyl and aryl derivatives, M—R, the hydrocarbon group is regarded as the radical R· bonded to the metal M·. Likewise it is more convenient, though less conventional, to regard covalent metal-halogen bonds as being derived from the halogen radical, e.g. Cl·, rather than from addition of, say, Cl⁻ to the metal, which relative to the anion must then have a unipositive charge, M^+.

150

Table XI. *A classification of organic groups which act as ligands to transition metals*

Number of electrons	Name of class	Examples of organic group
1	Yl	Alkyl or aryl groups, compare Cl·, ·CN and H·
2	Ene	Ethylene, compare carbon monoxide
3	Enyl	π-Allyl, compare nitric oxide
4	Diene	Cyclobutadiene, butadiene
5	Dienyl	π-Cyclopentadienyl
6	Triene	Benzene, cycloheptatriene
7	Trienyl	π-Cycloheptatrienyl
8	Tetraene	Cyclo-octatetraene

18 electrons gives rise to stable compounds. Exceptions to the rule are discussed later.

In order to test the application of the 18-electron rule to a compound it is necessary to count the electrons in the valence shell of the metal, which can be done as follows using Table XI:

(*a*) Take the number of electrons in the valence shell of the uncomplexed metal atom and add or subtract electrons according to the total charge on the metal complex. Thus iron in a unipositive cation contributes $8-1 = 7$ electrons, in a neutral compound 8 electrons and in a uninegative anion 9 electrons.

(*b*) Sum the number of electrons which the ligands formally contribute to the metal, according to the classification of ligands given above. Add this sum to the number of metal electrons from (*a*).

To exemplify this procedure consider first some simple carbonyl compounds:

Metal Carbonyl	Number of electrons contributed by metal	Number of electrons contributed by ligands	Total
$Fe(CO)_5$	8	$2 \times 5 = 10$	18
$Mo(CO)_6$	6	$2 \times 6 = 12$	18
$[Mn(CO)_6]^+$	$7-1=6$	$2 \times 6 = 12$	18
$[Co(CO)_4]^-$	$9+1=10$	$2 \times 4 = 8$	18
$V(CO)_6$*	5	$2 \times 6 = 12$	17

*Paramagnetic.

In those compounds where metal-metal bonds are present, for example $(OC)_5Mn-Mn(CO)_5$, each metal is regarded as acting as a one electron ligand M· to the other. Thus the number of electrons about a manganese atom in dimanganese decacarbonyl is $7+(2\times5)+1 = 18$.

Consider now examples of organometallic complexes:

Compound	Number of electrons contributed by metal	Number of electrons contributed by ligands	Total
$(\pi\text{-}C_5H_5)_2Fe$	8	2×5 $(\pi\text{-}C_5H_5)_2$	18
$\pi\text{-}C_5H_5Mo(CO)_2\pi\text{-}C_3H_5$	6	5 $(\pi\text{-}C_5H_5)$ 2×2 $(CO)_2$ 3 $(\pi\text{-}C_3H_5)$	18
$[\pi\text{-}C_5H_5Fe(CO)_2C_2H_4]^+$	$8-1=7$	5 $(\pi\text{-}C_5H_5)$ 2×2 $(CO)_2$ 2 (C_2H_4)	18
$\pi\text{-}C_5H_5Fe(CO)PPh_3Me$	8	5 $(\pi\text{-}C_5H_5)$ 2 (CO) 2 (PPh_3) 1 (Me)	18
$(Et_3P)_2PtClC_6H_5$	10	2×2 $(Et_3P)_2$ 1 (Cl) 1 (C_6H_5)	16

In some molecules the molecular formula provides insufficient information with which to test the application of the 18-electron rule, and further knowledge, for example of the structure, is required. At first sight the compound $\pi\text{-}C_5H_5Re(CO)_2C_5H_6$ appears to disobey the rule (Re = 7, $\pi\text{-}C_5H_5 = 5$, $(CO)_2 = 4$, C_5H_6 as a diene = 4, the total is 20 electrons.) It is found, however, that the compound is readily reduced by hydrogen to give the cyclopentene compound, 5.2. This suggests that in the parent compound only one of the double bonds of the cyclopentadiene ligand is attached to the rhenium and that the compound has the structure, 5.1.

The structure of a complex may sometimes be *predicted* on the basis of the 18-electron rule. For example the cobalt atom in $C_7H_7Co(CO)_3$ appears to possess a 22-electron configuration (Co = 9, $C_7H_7 = 7$,

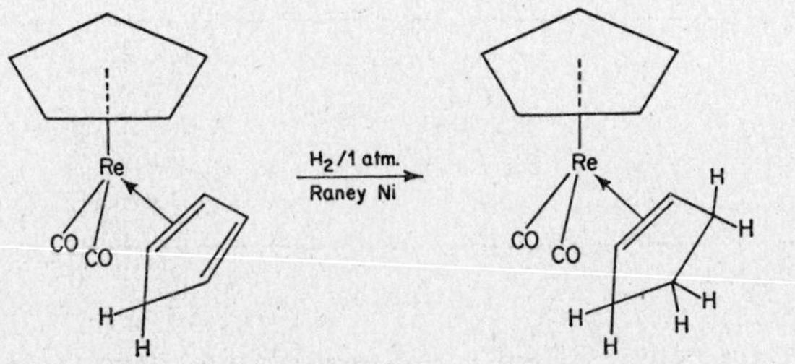

$H_2/1$ atm.
Raney Ni

5.1   5.2

$(CO)_3 = 6$). It is therefore *proposed* that the C_7H_7 ring contributes only 3, rather than 7 electrons to the cobalt, that is, the compound is not a π-cycloheptatrienyl complex, 5.3, but a π-enyl complex, 5.4.

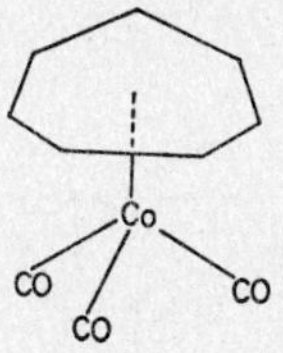

5.3 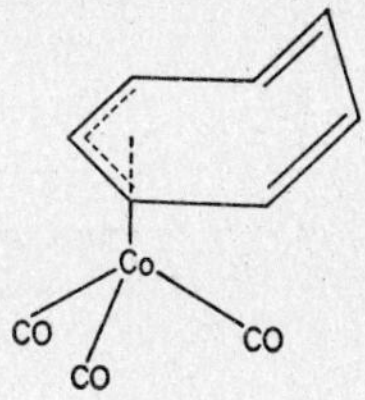   5.4

Formal oxidation state

It should be noted that the above procedure of counting valence electrons takes no account whatsoever of the concept of formal oxidation state. It is indeed rather confusing to try simultaneously to count electrons and to ascribe a formal oxidation state to the metal. While the concept of formal oxidation state is quite useful in the traditional classification of classical inorganic complexes, it is generally less helpful in discussing organo-metallic and related compounds. For example in the anion $[ReH_9]^{2-}$, is the rhenium in the $Re^{7+} + (H^-)_9$, in the $Re^{11-} + (H^+)_9$ or in the $Re^{2-} + (H\cdot)_9$ state?*

Since the *actual* charge on the metal is rarely larger than ± 1 (this is the Pauling Electroneutrality Principle), the unreality of assigning a formal charge far beyond these limits is clear.

The applicability of the 18-electron rule

The 18-electron rule applies most generally when all the metal ns, np and $(n-1)d$ orbitals are in the valence band. Figure 25 shows that for the first row transition metals this is nearest to the truth for the elements in the middle of the transition series (V, Cr, Mn, Fe, Co). For Sc and Ti the $3d$ orbitals are rather high in energy compared with the $4s$ and $4p$, whilst in Ni, Cu and especially in Zn the $3d$ orbitals have begun to enter the core and hence can be expected to make less than their full contribution to metal-ligand bonding. Also across the series Co, Ni, Cu and Zn the energy separation between the $4s$ and $4p$ orbitals increases. The stepwise removal of two of the three $4p$ orbitals may account for the occurrence of stable complexes of Co, Ni and Cu in which the metals have 18-, 16- and 14-electron environments respectively.

*For further discussion see Nyholm and Tobe, *Advances Inorg. Radiochem*, 1963, **5**, 1.

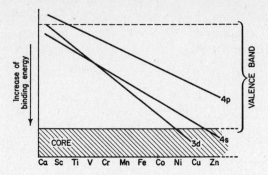

Figure 25. The change in energy of the 3*d*, 4*s* and 4*p* orbitals of the first transition series (After Phillips and Williams).

Basis of the 18-electron rule

The 18-electron rule arises mainly from the kinetic stability which may be associated with filled electron shells. Maximum use of valence orbitals also implies a maximum number of *bonding* electrons. Figure 26 (a) shows a qualitative molecular orbital diagram for an octahedral complex of the transition elements, considering only σ-bonding between metal and ligands. The relative positions of orbitals within the sets of bonding or of anti-bonding M.O.s are uncertain. In an octahedral environment the e_g (d_{z^2} and $d_{x^2-y^2}$) orbitals of the metal which lie along the axes and hence are directed towards the ligands form σ-bonding and σ-antibonding M.O.s *do not interact* with ligand orbitals of suitable symmetry, whereas the t_{2g} (d_{xy}, d_{yz} and d_{zx}) metal orbitals are non-bonding in this situation. The metal 4*s* and 4*p* orbitals can also form σ-M.O.s with the ligand orbitals.

From Figure 26(a), it can be seen that 18 electrons in all are required to fill all the bonding and non-bonding molecular orbitals in the complex. These electrons come both from the metal and from the ligands. Kinetic stability of the complex will be achieved if no low-lying orbitals are available into which electrons may readily be promoted to initiate thermal decomposition, or donated, as in nucleophilic attack. It is clear that kinetic stability is unlikely to be attained if bonding or non-bonding M.O.s are empty, that is, if less than a total of 18 electrons is present. This is the molecular orbital explanation of the 18-electron rule.

Even when this environment of 18 electrons is present, a compound will be thermally stable (e.g. survive at room temperature) only where the energy gap between the highest filled and lowest unfilled M.O.s is sufficiently large. This energy gap (Δ) may be associated with the enthalpy contribution to the free energy of activation (E_A) shown in Figure 5, p 9.

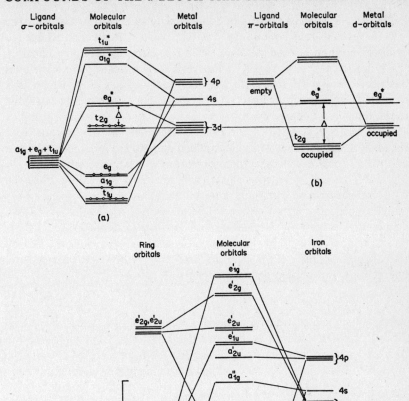

Figure 26 (a) A qualitative molecular orbital diagram for an octahedral complex ML_6, showing only those interactions between metal-atomic and ligand orbitals which lead to metal-ligand bonds of σ-symmetry.

(b) Shows how the energy separation Δ between the t_{2g} and $e_g{}^*$ molecular orbitals may be increased when π-'back-donation' from metal to ligands occurs.

(c) A molecular orbital diagram for ferrocene (After Schustorovich and Dyatkina, and Cotton).

(i) The role of π-bonding ligands. The majority of complexes which obey the 18-electron rule contain ligands (e.g. CO, R_3P, olefins) which are thought to π-bond to the metal and to *accept* electrons from the metal by 'back-donation' (see p 186). For this reason such ligands are sometimes described as π-acids (compare the acceptor molecule BF_3 which is a σ-acid). These π-acceptor ligands have two functions. First they can assist in the removal of charge from the metal atom. In other words a complex such as $Mo(NH_3)_6$ would be unstable because the twelve electrons from the ammonia lone pairs would be confined close to the molybdenum atom and would repel each other. In contrast, in the relatively stable carbonyl $Mo(CO)_6$ the metal electrons can be delocalized by π-bonding on to the unsaturated carbonyl groups and a large build-up of electron density on the molybdenum atom is thus prevented. Secondly, it is argued that π-bonding ligands may in certain circumstances increase the energy gap Δ. This is illustrated in Figure 26 (b). The t_{2g} orbitals in a complex where the ligands form only σ-bonds are non-bonding, and may be lowered in energy by interaction with unfilled ligand π-orbitals.

(ii) Comparison of 3d, 4d and 5d transition series. From studies of the electronic spectra of transition metal complexes it has been shown that, for analogous complexes, Δ increases by about 30% from the 1st to the 2nd, and again from the 2nd to the 3rd transition series. Qualitative measures of 'stability', e.g. decomposition temperatures, seem to support the hypothesis that, in some cases at least, organometallic compounds of the 3rd transition series are more stable than the corresponding complexes of the 1st or 2nd transition series elements.

(iii) Exceptions to the 18-electron rule.

(a) 16-electron (d^8) and d^3 complexes. These electron configurations are also found for thermally stable complexes containing metal-carbon bonds. An important class of 16-electron complexes are those formed by d^8 metal ions, e.g., Rh(I), Ir(I), Ni(II), Pd(II), Pt(II), Au(III). The complexes have a square planar distribution of ligands about the metal atom.

Octahedral metal complexes of d^3, Cr(III) and low spin d^6, Co(III),* have particularly high kinetic stability to ligand replacement or exchange. The activation energy for ligand substitution in such complexes has been shown to be considerably greater than in comparable complexes in which the metal possesses other d^n configurations. This kinetic stability is associated with the half filled or filled t_{2g} environments, in which the

*All known octahedral cobalt alkyls in fact obey the 18-electron rule. It is, however, relevant to consider them together with the Cr(III), d^3 complexes.

electron density is essentially spherically symmetrical about the metal atom, so that there is no favoured path for attack by polar nucleophiles. It has also been correlated with the crystal field interaction energy.

Important examples of octahedral d^3 and d^6 complexes are Vitamin B_{12} and certain alkyl and aryl derivatives of Cr(III) and Co(III) (see p 243). (*b*) *Stabilization by steric effects.* There is a number of thermally quite stable compounds which do not obey the 18-electron rule, for the stability of which explanations based on steric effects may be advanced. It is probable that the paramagnetic, monomeric $V(CO)_6$ does not dimerise because this would require a co-ordination about the vanadium of seven (or more), one of the ligands being the large $V(CO)_6$ group.

The 'ortho effect'—namely the thermal stabilization of ortho-substituted aryl complexes, e.g. $(R_3P)_2Ni(Mesityl)_2$ (see p 185)—may be due, in part, to steric inhibition by ligands of attack along the z-axis (see Figure 36, p 186).

Bonding in organometallic π-complexes

The ethylene-metal system provides a useful model by which to describe the bonding in π-bonded metal-organic complexes. As we shall see, such bonding is difficult to treat in terms of a conventional classical valence bond approach. It is more meaningful to use descriptions based on molecular orbital theory.

Consideration of the symmetry of a molecule followed by the application of group theory enables us to determine which orbitals of the metal and of the ligands are allowed to combine to form molecular orbitals, though a significant degree of chemical bonding results only when they are of suitable and similar energy. In the following discussion we shall be concerned chiefly with one type of symmetry element—a rotational axis of symmetry perpendicular to the plane containing the organic ligand which is being considered and passing through the metal atom. This axis is by convention called the z-axis. Using this axis, the ligand and metal orbitals may frequently be classified as σ-, π-, or δ-orbitals (provided that the symmetry of rotation of the molecule is > 2). (See Table XII).

The wave-function of a σ-orbital does not change sign on rotation through 180° about the axis of symmetry (z-axis). For a π-orbital, the sign of the wave-function changes once, and for a δ-orbital twice on such rotation. This classification is further illustrated in Figure 31, p 164.

Since only those ligand and metal orbitals which have the same symmetry properties can combine to form bonding M.O.s, we may by symmetry considerations determine the often fairly small number of combinations of ligand and metal orbitals which can lead to bond formation.

Table XII. *Symmetry classification of metal orbitals with respect to rotation about the z-axis*

Symmetry	Metal orbitals
σ	s, d_{z^2}, p_z
π	d_{zx}, d_{yz}, p_y, p_x
δ	$d_{x^2-y^2}, d_{xy}$

It should be noted that we are restricting ourselves to the valence atomic orbitals of the metal and the $p\pi$-M.O.s of the ligand.

Energies of molecular orbitals

Symmetry considerations alone tell us nothing about the energies of the molecular orbitals derived from the metal and ligand orbitals, or even about the relative order in which they lie. It can be stated, however, that a large overlap between orbitals of similar energy leads to strong bonding, and that the closer in energy the interacting orbitals are, the more stable the resulting bonding molecular orbital will be. Experimental evidence about the relative energies of the molecular orbitals can often be obtained from spectroscopic (visible and ultra-violet) and magnetic data. Such energies are normally expressed in terms of eV or kcal/mole below the onization energy.

A description of the bonding of ethylene to transition metals

Consider the bonding of ethylene in Zeise's salt $K^+[C_2H_4PtCl_3]^-$. X-ray diffraction studies have shown that the anion has the structure given in Figure 27. The platinum is in a square planar configuration with the axis

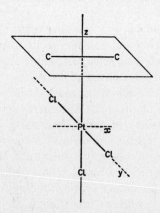

Figure 27. Structure of the anion $[C_2H_4PtCl_3]^-$.

of the C=C double bond lying perpendicular to this plane. The anion has C_{2v} symmetry; the z-axis is drawn according to convention, coincident with the 2-fold rotation axis of symmetry. In the discussion which follows, the electrons which are involved in the σ-bonding of the H_2C-CH_2 system are assumed *not* to be involved in the bonding to the metal. The ethylene-metal bond is considered to arise from interaction between the $2p_z$ ($2p\pi$) orbitals of the carbon atoms and metal orbitals of suitable symmetry. It must be emphasized, however, that the description of the internal bonding of the ethylene unit as involving carbon sp^2 hybridization (as is true for the free ethylene molecule) is often only a crude approximation in metal-ethylene complexes (see p 162).

The various interactions between metal and ethylene p_z orbitals which are permitted by symmetry considerations are represented pictorially in Figure 28.

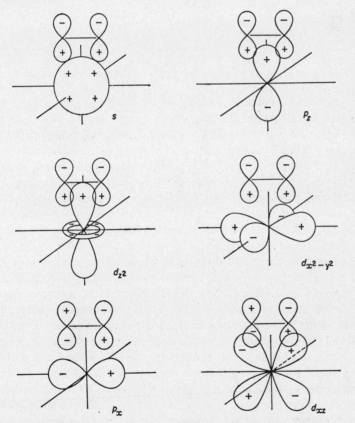

Figure 28. Diagrammatic representation of types of orbital overlap between ethylene $2p_z$ atomic orbitals and ns, np and (n−1)d orbitals of a transition metal.

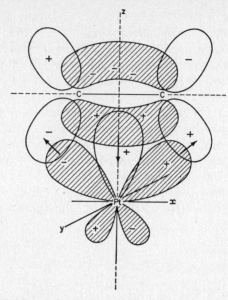

Figure 29. Conventional representation of the metal-olefin bond.

Another traditional representation of the metal-olefin bond is given in Figure 29. In this picture the two carbon p_z orbitals have first been combined to form a π-bonding and a π-antibonding molecular orbital, and the metal orbitals have been hybridized. One of the metal hybrid orbitals is able to overlap with the filled bonding M.O. of the ethylene to give a σ-bond, while the other can interact with the empty antibonding M.O. of the ligand, which results in a bond with π-symmetry about the z-axis. Thus the metal-olefin bond consists of two components, σ-donation from the ligand to the metal, and simultaneous π-donation from the metal to the ligand. This description of the bonding is very similar to that given for the metal-carbon monoxide bond in metal carbonyls (see p 186). Both ethylene and carbon monoxide are too weak σ-donors to form adducts with typical Lewis acids such as Me_3B, which form strong complexes with Lewis bases such as ammonia. Unlike ammonia, however, they possess vacant low-lying orbitals which can accept electrons from filled or partially filled d-orbitals of transition elements. The transfer of electrons from the ligand to the metal in the formation of a σ-bond can probably be enhanced by the simultaneous removal of charge from the metal through π-back-donation to the ligand (a 'synergic' effect). It is difficult to assess the relative importance of σ- and π-bonding in such systems. There is evidence to suggest, however, that in general π-bonding makes the greater contribution, especially through use of a metal $d\pi$-orbital.

As π-donation from metal to ligand takes place from filled metal orbitals, olefin-metal complexes are usually formed by metals in low, electron-rich, oxidation states, especially by those elements late in the transition series such as Fe, Co, Ni and Cu, and their congeners in later Periods. These elements of course have more valence electrons than the earlier ones.

Both removal of electrons from the bonding orbital of ethylene, and donation into its π-antibonding orbital will tend to weaken the C—C bond. In accordance with this a lowering of the C=C stretching frequency of 60–150 cm^{-1} is usually observed in olefins co-ordinated to transition metals. For example a band at 1511 cm^{-1} in the infrared spectrum of Zeise's salt has been assigned to the C=C stretch as compared with a frequency of 1623 cm^{-1} in the Raman spectrum of ethylene.

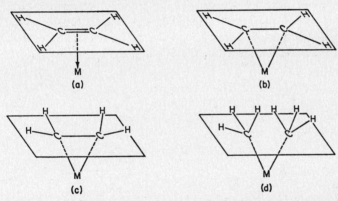

Figure 30. Valence bond representations of the ethylene-metal bond (except *d*, see below).

Figure 30 (*a*), (*b*) and (*c*) gives representations of the bonding in valence bond terms, but these are rather limited descriptions. Such descriptions are useful in emphasizing the relative importance of the contributions to the metal-ligand bond of σ-bonding (*a*) and π-bonding (*b*). This problem of representation is further illustrated by the rather extreme example of cyclopropene, which is usually written as 5.5, but an alternative 'representation' is 5.6, where the molecule is shown as a π-acetylene complex of carbene.

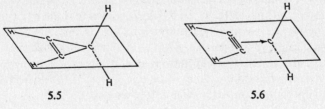

Evidence of the relative contributions of the various valence bond structures (or of σ- and π-bonding) can sometimes be obtained from structural determinations. If the configuration around the carbon atoms of a co-ordinated ethylene molecule were found to be essentially tetrahedral, (indicating sp^3 hybridization), the ligand-metal bond could be thought of as being equivalent to two metal σ-alkyl bonds (Figure 30c), compare the dimethyl compound (Figure 30d). (See also p 226). In such a case (in M.O. language) the contribution of π-back-donation from filled metal orbitals (p_x, d_{zx}) into the π-antibonding orbitals of the olefin would be very important indeed, (Figure 29). The π-acceptor character of the ligand will clearly be enhanced when electron-attracting substituents are attached to carbon. Simultaneously its σ-donor character will be reduced. The crystal structure of the fumaric acid iron tetra-carbonyl complex, 5.7, shows that the carboxyl groups are bent out of the C=C axis, away from the $Fe(CO)_4$ group. Thus the C—COOH bond shows a little greater p-character than sp^2 and there is probably considerable back-donation from the $Fe(CO)_4$ group to the $p\pi$-orbitals of the olefinic system.

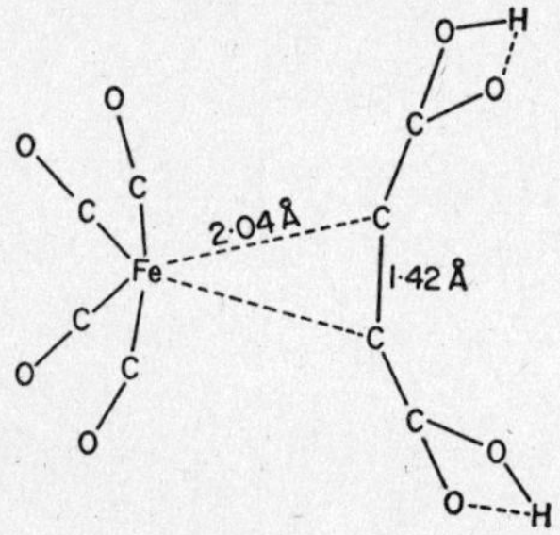

5.7

The rotation of ethylene about the ligand-metal bond

The ethylene-metal bond has been represented above as consisting of a σ- and a π-component. It therefore has some similarity to the C=C double bond in ethylene itself. Rotation about a carbon-carbon double bond is severely restricted, and stable geometrical isomers can be isolated. Some elegant variable-temperature proton magnetic resonance studies on the complex π-$C_5H_5Rh(C_2H_4)_2$, 5.8, have shown that the ethylene groups rotate about the metal-ethylene axis with a rotational energy barrier of $\sim$6kcal. The ethylene ligands do not exchange readily with free ethylene, however, and it therefore appears that ethylene has a choice of metal orbitals with which to combine, so that as it rotates it is always π-bonded to the metal, although some orientations are favoured.

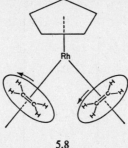

5.8

For similar reasons, π-C_4, π-C_5 and π-C_6 rings are able to rotate quite readily about the metal-ligand axis. For example the energy barrier to rotation of the π-cyclopentadienyl ligands in ferrocene is about 1 kcal./mole.

The bonding of other unsaturated hydrocarbons to transition metals

The bonding of other hydrocarbon ligands to transition metals can be treated by a similar molecular orbital approach. The electrons involved in the internal σ-bonding of the carbon and hydrogen atoms of the unsaturated hydrocarbon group are again assumed, to a first approximation, to play no part in the bonding of this group to the metal. After the skeletal σ-bonds have been formed, one $2p_z$ orbital remains on each carbon atom which can interact with the other $2p_z$ orbitals on the other carbon atoms of the delocalized system to form an equal number of π-molecular orbitals.

These molecular orbitals, which are illustrated in Figure 31 for a series of hydrocarbon ligands, may be classified as having σ-, π-, or δ-symmetry with respect to rotation about the z-axis. Clearly these ligand orbitals can interact only with those metal orbitals which possess similar symmetry properties. This is further illustrated by a molecular orbital diagram for ferrocene $(\pi$-$C_5H_5)_2$Fe (Figure 26c). While the relative orders of the energy levels are not yet decided, it is generally agreed that, as in the case of the ethylene-metal bond, the major contribution to the bonding results from π-overlap between metal orbitals of π-symmetry and the ligand M.O.s ψ_2 and ψ_3.

The 18 available electrons in ferrocene (ten from the two C_5H_5 ligands and eight from the iron atom) may be placed into the nine M.Os. of lowest energy.

As the 3-electron π-allyl- and the 4-electron π-butadiene-metal systems have no elements of rotational symmetry about the z-axis, a very large

L

	Classified as σ—symmetry	Classified as π—symmetry	Classified as δ—symmetry
3—electron π—allyl	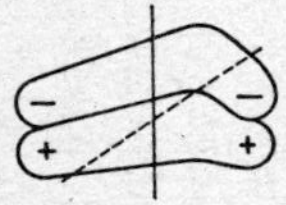	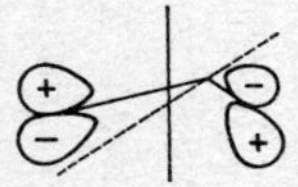	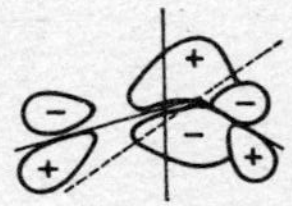
4—electron cyclobutadiene	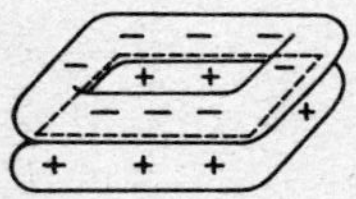	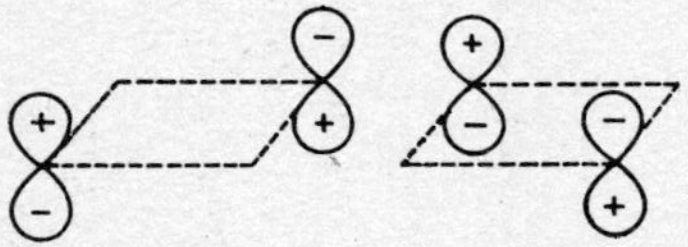	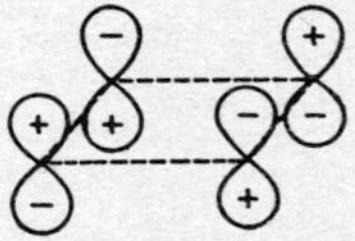
butadiene	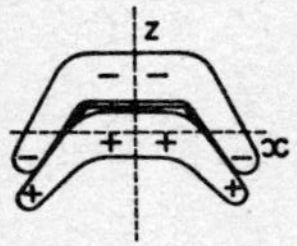	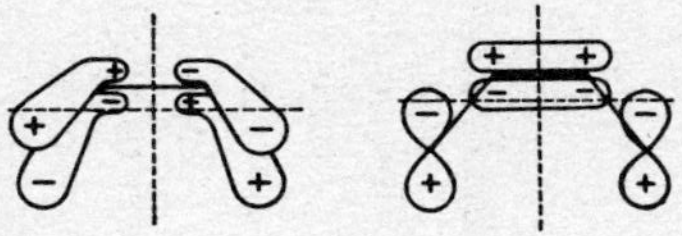	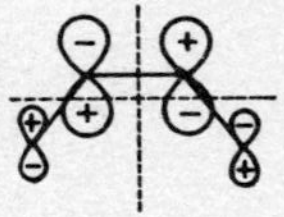
5—electron π—cyclopentadienyl	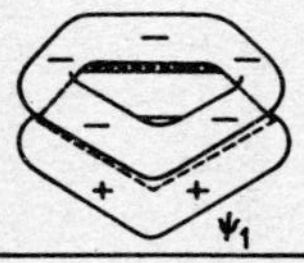 ψ_1	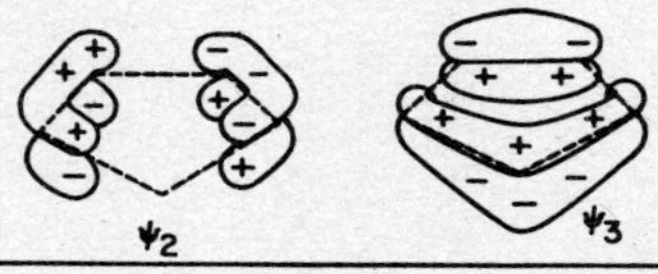 ψ_2 — ψ_3	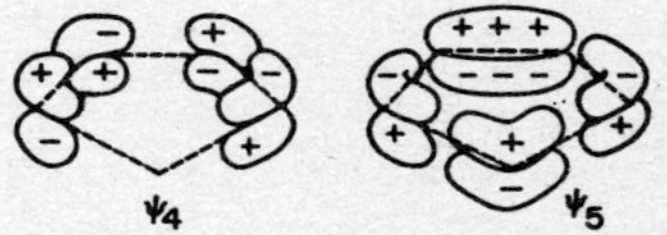 ψ_4 — ψ_5

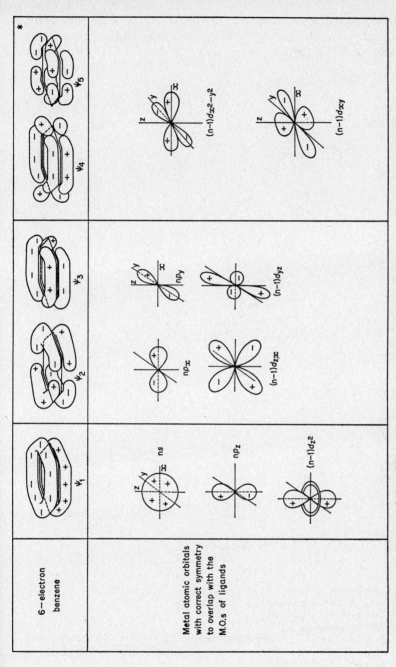

Figure 31. Representation of ligand π-molecular orbitals and their possible interactions with metal orbitals; for simplicity the radial nodes of metal orbitals are not shown.

φIn the cases of π-allyl and butadiene, this symmetry classification is incorrect, see p 163, 166.

*The highest energy benzene m.o, (ψ_6), is omitted as it will make either little or no contribution to bonding.

165

number of interactions between metal and ligand orbitals is allowed. Of the interactions in Figure 31, only those in which significant overlap occurs are likely to be important. They may be classified for consistency

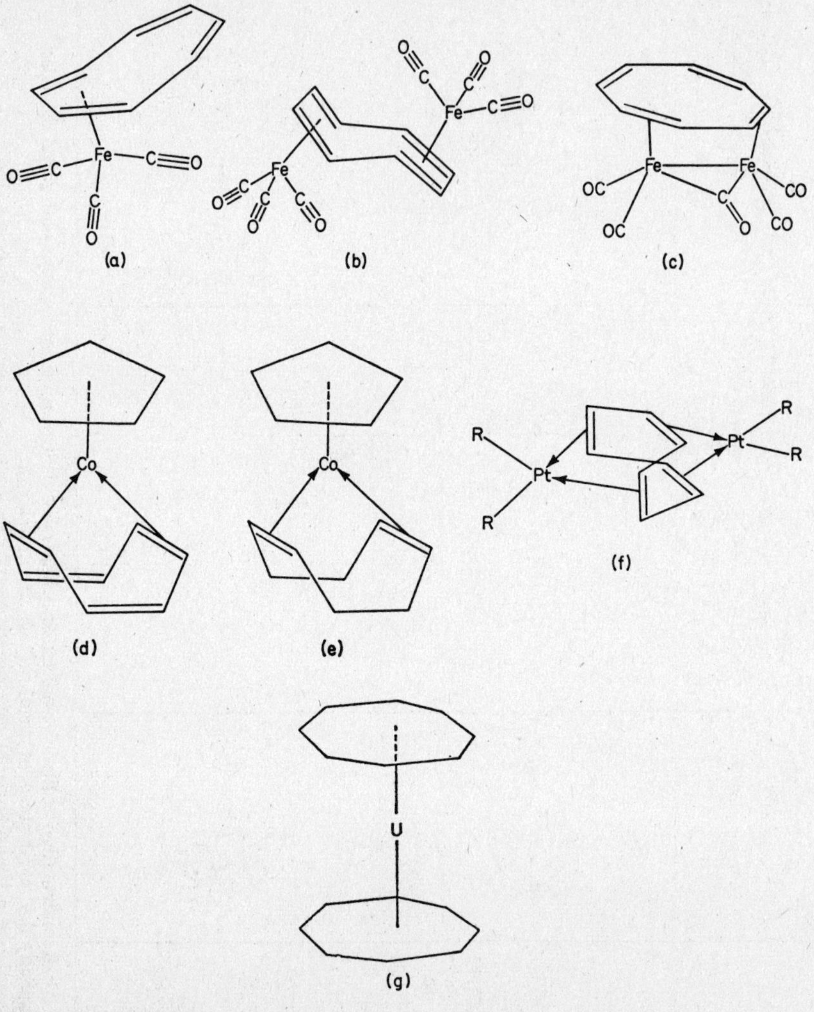

Figure 32. Structures of some cyclo-octatetraene and related complexes.
    (a) cyclo-octatetraene acting as a 4-electron ligand;
    (b) as a 2×4-electron ligand;
    (c) as a 2×2-electron ligand;
    (d) as a chelating 2×2-electron ligand, compare the cyclo-octa-1,5-diene complex (e);
    (f) as a 4×2-electron ligand;
    (g) as an 8-electron ligand.

by comparison with the cyclic C_nH_n systems as giving rise to mainly σ-, π- and δ-type bonds.

The bonding of all unsaturated hydrocarbon ligands to transition metals thus consists of two or more parts—

(*a*) σ-donation from filled bonding M.O.s on the ligand into empty metal orbitals of σ-symmetry.

(*b*) π-interaction (the ligand orbitals may be filled as in benzene, or partly filled as in cyclobutadiene, so that they may have donor or acceptor properties).

(*c*) δ-donation from the filled metal $d_{x^2-y^2}$ or d_{xy} orbital into empty antibonding orbitals on the ligand. This is likely to be of minor importance owing to the high energy of these empty ligand orbitals.

Comparison of conjugated and non-conjugated olefins as ligands

Unconjugated olefins such as cyclo-octa-1,5-diene can be considered to bond to transition metals essentially by isolated 2-electron olefin-metal linkages. On the other hand the bonding of conjugated olefins such as butadiene is best treated in terms of their delocalized π-M.O.s. For this reason it is useful to distinguish between conjugated and non-conjugated olefins as ligands. In fact an olefin may act as a ligand in many different ways; this is well illustrated by the structures of some cyclo-octatetraene complexes in Figure 32.

The example of cyclo-octatetraene emphasizes that all the carbon-carbon double bonds in a di- or poly-olefin are not necessarily used in complex formation. Other examples are found elsewhere in this book— butadieneFe(CO)$_4$ (p 172) and cycloheptatrieneFe(CO)$_3$ (p 197).

BIBLIOGRAPHY

D. A. Brown, in *Transition Metal Chemistry*, Volume 3, edited by R. L. Carlin, p. 1. (Edward Arnold, London, 1966). Electronic structure of some organometallic molecules. An advanced review.

F. A. Cotton, 'Chemical Applications of Group Theory' (Interscience, New York, 1963). A good introductory text.

F. A. Cotton and G. Wilkinson, 'Advanced Inorganic Chemistry', 2nd Edition (Interscience, New York, 1966) chapters 25, 26, 27.

C. A. Coulson, 'Valence', 2nd Edition (Oxford University Press, Oxford, 1961).

L. E. Orgel, 'An Introduction to Transition Metal Chemistry, Ligand Field Theory' (Methuen, London, 1960).

C. S. G. Phillips and R. J. P. Williams, 'Inorganic Chemistry', (Oxford University Press, Oxford, 1966). Chapters 24, 27.

J. W. Richardson, in 'Organometallic Chemistry', edited by H. Zeiss, (Reinhold, New York, 1960). Carbon-metal bonding. An examination of fundamental principles.

CHAPTER SIX

Preparation of organo-transition metal compounds

General considerations

In Chapter 1 it was pointed out that the thermal stability of many organo-metallic compounds of the transition elements is rather low. Whereas a few compounds, for example ferrocene $(\pi\text{-}C_5H_5)_2Fe$ and dibenzene chromium $(\pi\text{-}C_6H_6)_2Cr$ can be heated unchanged *in vacuo* to temperatures exceeding 250°C, the majority decompose at temperatures in the range 100–200°C, and some at or below room temperature. It is therefore necessary to bear in mind the possible thermal instability of reactants and products and to use as mild conditions of temperature as are feasible.

It was also noted in Chapter 1 that all organometallic compounds are thermodynamically unstable to attack by oxygen and that in many cases, oxidation occurs at a significant rate at room temperature. Sensitivity to oxygen is usually greater in solution than in the solid state, so that the absence of oxygen is obligatory while carrying out reactions involving most organometallic compounds. This is achieved either by working under vacuum or in an inert atmosphere such as nitrogen or argon. Most operations involved in the preparation of ordinary organometallic compounds can be carried out under an inert atmosphere with little difficulty. Clearly great care is required in choosing experimental conditions in this field.

Solvents. Most reactions proceed most readily under homogeneous conditions, and where fairly low temperatures are necessary, as in organometallic chemistry, this means that solution techniques are commonly used. Generally organic solvents such as ethers (diethyl ether, tetrahydrofuran), aromatic hydrocarbons and saturated hydrocarbons are most suitable.

Starting materials. As starting materials metal compounds which are soluble in suitable organic solvents and which are readily available, either

168

commercially or by simple preparations, are required. Most commonly anhydrous metal halides or other neutral complexes such as acetylacetonates or carbonyls are suitable.

Preparative routes

Some important routes are described below. Examples are chosen to illustrate both the generality of the method and important classes of complexes.

From metal salt, reducing agent and ligand. When the metal in the metal halide is in a higher formal oxidation state than in the organometallic compound which is being prepared, it is necessary to add a reducing agent to the reaction mixture. This is a general principle in the preparation of all 'low oxidation state' complexes, for example metal carbonyls, olefin complexes, arene complexes, molecular nitrogen complexes, etc.

Some examples of the application of this method are listed in Table XIII.

Table XIII

Metal Salt	Reducing Agent	Ligand	Product
$MnCl_2$	$Ph_2CO^-Na^+$	CO/100 atm. 150°C	$Mn_2(CO)_{10}$
$CrCl_3$	Zn/Hg	CO	$Cr(CO)_6$
$CrCl_3$	Al (AlCl₃)	Benzene	$(\pi\text{-}C_6H_6)_2Cr^+$
$FeCl_3$	RCH_2CH_2MgX	Olefin	Fe° olefin complex
$NiCl_2$	R_3Al		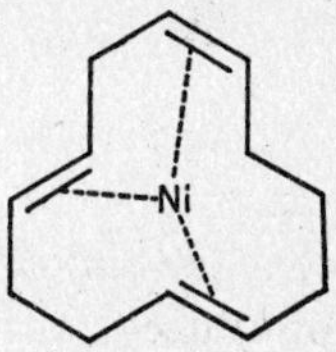
$(Ph_3P)_2PtCl_2$	$N_2H_4.H_2O$	Olefin	$OlefinPt(PPh_3)_2$

The reduction of a metal salt by aluminium and aluminium trichloride in the presence of an aromatic hydrocarbon, which is illustrated in Table XIII, provides the most general method known for the preparation of bis-π-arene complexes. It was first used by Fischer and Hafner for the synthesis of bis-π-arene chromium complexes (Fischer's reducing Friedel-Crafts procedure).

$$3CrCl_3 + Al + 2AlCl_3 + 6C_6H_6 \longrightarrow 3(\pi\text{-}C_6H_6)_2Cr^+AlCl_4^-$$
6.1

The cation 6.1 may readily be reduced to bis-π-benzene chromium by aqueous sodium dithionite:

$$2(\pi\text{-}C_6H_6)_2Cr^+ + S_2O_4{}^{2-} + 2OH^- \longrightarrow 2(\pi\text{-}C_6H_6)_2Cr + 2HSO_3{}^-$$

Alternatively, the neutral complex is obtained by the disproportionation of the bis-π-benzene chromium cation in aqueous alkaline solution

$$2(\pi\text{-}C_6H_6)_2Cr^+ \xrightarrow{\;H_2O\;} (\pi\text{-}C_6H_6)_2Cr + 2C_6H_6 + Cr^{2+}$$

Arene complexes of many of the d-block transition metals may be prepared similarly. As with chromium, the initial products of the reducing Friedel-Crafts procedure are usually cationic; the neutral bis-π-arene complexes are often obtained from them by reduction or by hydrolytic disproportionation.

The formation of cyclododecatriene nickel from nickel(II) chloride and butadiene in the presence of trialkylaluminium compounds is an example of the oligomerization of a ligand during complex formation. Further examples of this will be found on pp 239–241.

From metal and ligand. When the metal in the starting material is already in a formal oxidation state suitable for the formation of the desired product, direct ligand substitution can occur, without the necessity to include a reducing agent in the reaction mixture. As a very simple example of this the direct formation of the carbonyls of iron and of nickel from carbon monoxide and the metal may be cited:

$$Fe^\circ + 5CO \longrightarrow Fe(CO)_5$$

$$Ni^\circ + 4CO \longrightarrow Ni(CO)_4$$

It is uncommon, however, for ligands to react directly with metals to yield complexes, probably as the products would be thermally unstable under the conditions, e.g. of high temperature, under which the reactions would occur at a reasonable rate.

From metal compound and ligand; ligand displacement. The treatment of solutions of many metal salts with olefins may yield a complex directly. For example, Zeise's salt, which was first isolated in 1827, is obtained as air-stable yellow crystals by bubbling ethylene through an aqueous solution of potassium tetrachloroplatinate(II),

$$K_2PtCl_4 + C_2H_4 \longrightarrow K^+[Pt(C_2H_4)Cl_3]^-.H_2O + KCl$$

Similarly aqueous solutions of silver nitrate or perchlorate absorb olefins:

$$Ag(H_2O)^+{}_n + \text{olefin} \rightleftharpoons Ag(H_2O)^+{}_{n-1}. \text{olefin} + H_2O$$

Aqueous rhodium (III) chloride and ethylene yield the complex, 6.2.

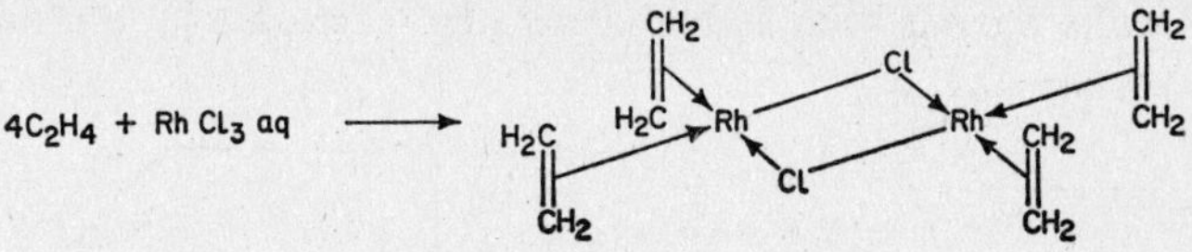

$$4C_2H_4 + Rh\,Cl_3\,aq \longrightarrow$$

6.2

Metal carbonyls are particularly convenient starting materials for the preparation of π-bonded hydrocarbon complexes. Carbon monoxide is replaced by the unsaturated hydrocarbon ligand. The volatility of the displaced carbon monoxide, which allows it to be removed readily from the reaction mixture, must assist the reaction.

Direct reaction between the ligand and the metal carbonyl frequently occurs at room temperature, e.g.

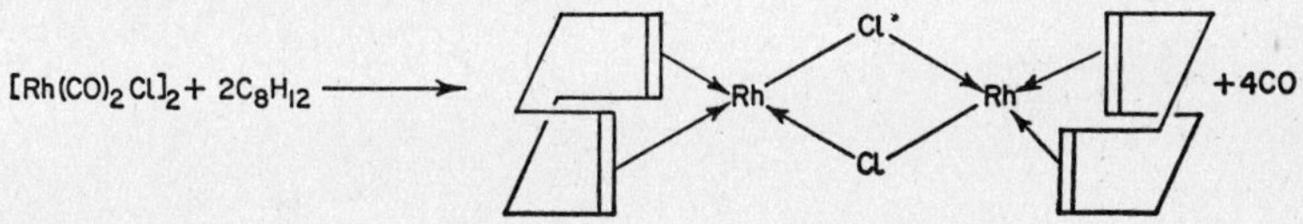

$$[Rh(CO)_2Cl]_2 + 2C_8H_{12} \longrightarrow \qquad +4CO$$

although sometimes more vigorous conditions are required e.g.

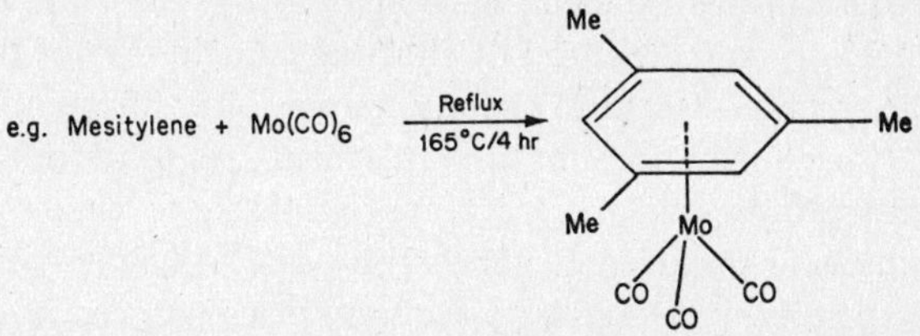

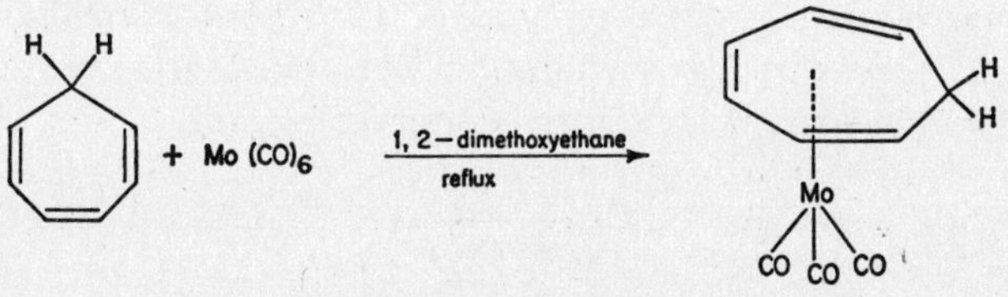

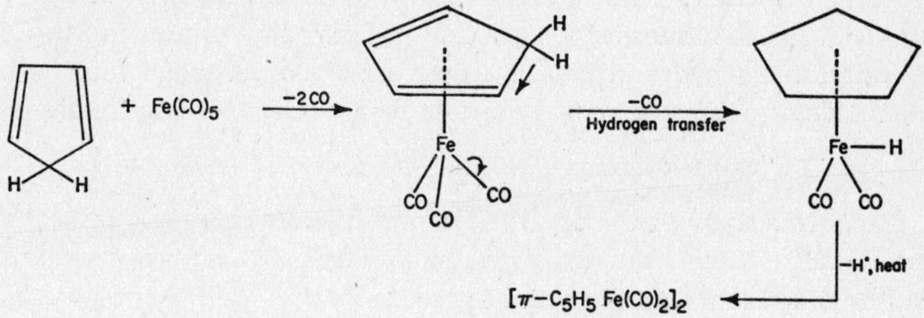

e.g. Mesitylene + $Mo(CO)_6$ $\xrightarrow[165°C/4\ hr]{Reflux}$

+ Mo (CO)$_6$ $\xrightarrow[reflux]{1,\,2-dimethoxyethane}$

+ Fe(CO)$_5$ $\xrightarrow{-2CO}$ $\xrightarrow[Hydrogen\ transfer]{-CO}$

$[\pi-C_5H_5\,Fe(CO)_2]_2$ $\longleftarrow$ $-H^{\cdot},\,heat$

With volatile ligands such as butadiene, the reaction may best be carried out under pressure, e.g.

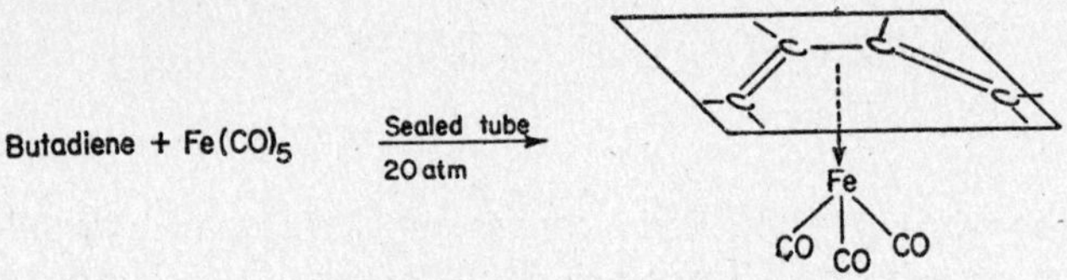

Butadiene + Fe(CO)$_5$ $\xrightarrow[\text{20 atm}]{\text{Sealed tube}}$

It is interesting to contrast the reaction of di-iron enneacarbonyl with butadiene at 40°C. Here a different product is formed, in which butadiene is acting as a 2-electron rather than as a 4-electron ligand.

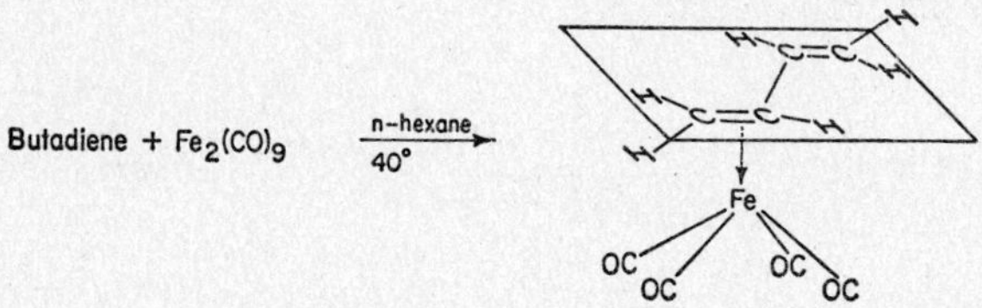

Butadiene + Fe$_2$(CO)$_9$ $\xrightarrow[\text{40°}]{\text{n--hexane}}$

Irradiation of solutions of metal carbonyls in the presence of an olefin is frequently a very good method of effecting substitution of the carbonyl ligands. Indeed photochemical techniques are often more advantageous than thermal reactions—any unstable intermediates produced are formed at lower temperatures and therefore are less liable to decompose to unwanted products. Thermal and photolytic reactions may yield different products. Generally, photochemical substitutions proceed by a dissociative (S$_N$1) mechanism whereas thermal reactions may proceed by either S$_N$1 or associative (S$_N$2) mechanisms. Thus photochemical techniques have been used to effect substitution when thermal substitution by an associative mechanism is difficult or impossible, e.g.

$$C_2H_4 + \pi\text{-}C_5H_5Mn(CO)_3 \xrightarrow[\text{petroleum}]{\text{u.v.}} \pi\text{-}C_5H_5Mn(CO)_2C_2H_4 + CO$$

Very rarely can all the carbonyl groups in metal carbonyls be replaced by olefin ligands. Substitution of 2, 3 or 4 carbonyl groups in Mo(CO)$_6$ by unsaturated hydrocarbons has been observed, although four carbon monoxides are replaced only by certain dienes under forcing conditions:

$$Mo(CO)_6 + \text{cyclohexa-1,3-diene} \longrightarrow (\text{diene})_2Mo(CO)_2 + 4CO$$

Metal tricarbonyl systems, M(CO)$_3$, are usually markedly resistant to further substitution; thus only one or two carbonyl groups may readily be displaced from iron pentacarbonyl.

Other ligands which are readily displaced by unsaturated hydrocarbon ligands include nitriles, e.g.

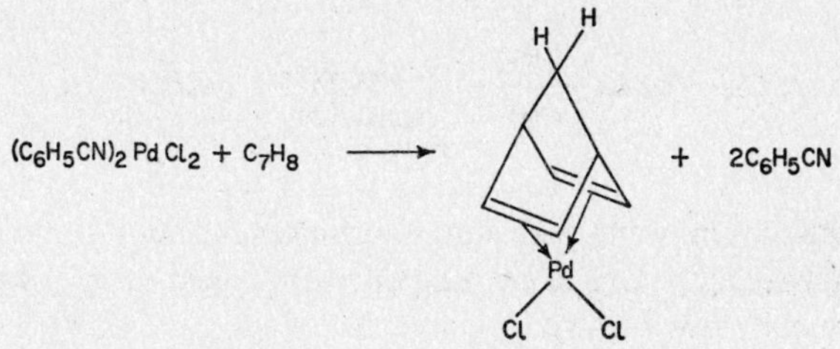

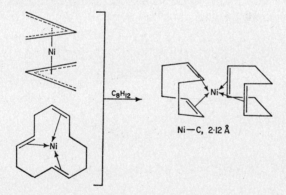

(C6H5CN)2 PdCl2 + C7H8
H
H
+ 2C6H5CN
Pd
Cl
Cl

or other hydrocarbons, e.g.

Cr
CO
CO
CO
C7H8
H
H
Cr
CO
CO
CO

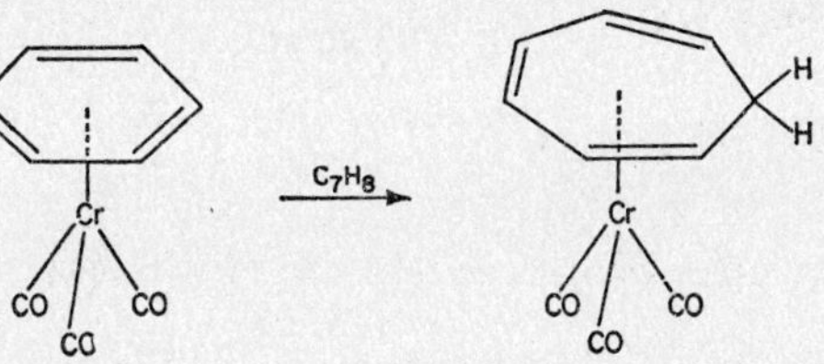

In particular, the ligands in bis-π-allyl nickel and in cyclododeca-1,5,9-triene nickel are readily displaced:

Ni
Ni
C8H12
Ni
Ni—C, 2·12 Å

Substitution by organometallic derivatives of main group elements:

As was shown in Chapter 2, organometallic compounds of the most electropositive elements, e.g. Li, Mg, react with halides of less electropositive main group elements by transfer of the organic group:

$$RM + M'Hal \longrightarrow MHal + RM'$$

Alkyls and aryls. This reaction also provides a general route to σ-alkyl and σ-aryl-transition metal complexes. A practical limitation is the stability of the organotransition metal derivative, e.g.

$$TiCl_4 + 4MeLi \xrightarrow[\text{ether}]{-80°C} \underset{\substack{\text{(unstable} \\ \text{above } -78°C)}}{Me_4Ti} + 4LiCl$$

As is shown below for some platinum complexes, organolithium reagents are more reactive than Grignard reagents, and yield fully alkylated or arylated products more readily:

$$\underset{\text{monoalkyl}}{\textit{trans-}(Et_3P)_2PtMeI} \xleftarrow{\ MeMgI\ } \textit{trans-}(Et_3P)_2PtCl_2 \xrightarrow{\ MeLi\ } \underset{\text{dialkyl}}{\textit{trans-}(Et_3P)_2PtMe_2}$$

$$\Big\downarrow RC{\equiv}CLi$$

$$\underset{\text{alkynyl}}{\textit{trans-}(Et_3P)_2Pt(C{\equiv}CR)_2}$$

Two further examples of the method are illustrated:

$$\pi\text{-}C_5H_5Fe(CO)_2R \xleftarrow{\ \pi\text{-}C_5H_5Fe(CO)_2Cl\ } RMgX \xrightarrow[\text{(R = Ph)}]{CrCl_3.3THF} Ph_3Cr.3THF$$

Allyls. Various transition metal halides yield allyl complexes when they are treated with allylmagnesium chloride, e.g.

$$MCl_2 + 2C_3H_5MgCl \longrightarrow \qquad (M = Ni, Pd, Pt)$$

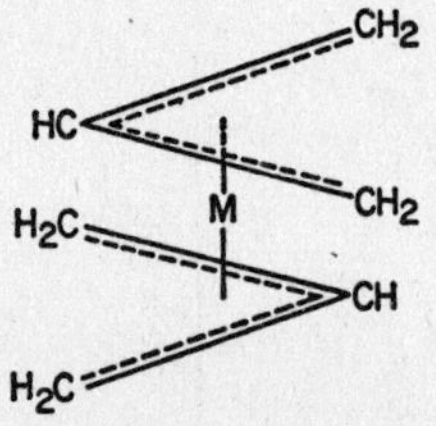

$$MCl_4 + 4C_3H_5MgCl \longrightarrow (\pi - C_3H_5)_4M \qquad (M = Zr, Hf)$$

Cyclopentadienyls. When anhydrous transition metal halides react with sodium cyclopentadienide in tetrahydrofuran, π-cyclopentadienyl complexes often result, e.g.

$$MCl_2 + 2NaC_5H_5 \xrightarrow[\text{reflux}]{THF} (\pi\text{-}C_5H_5)_2M + 2NaCl$$

$$(M = Fe, Co, Ni, Cr)$$

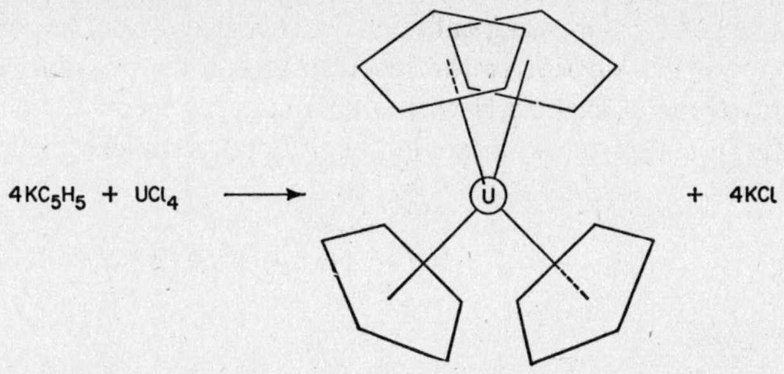

$$4KC_5H_5 + UCl_4 \longrightarrow \quad\quad + 4KCl$$

This is, indeed, the most important method of preparing π-cyclopentadienyls of the transition elements. Where the metal in the halide is in a higher formal oxidation state than in the product, sodium cyclopentadienide can act as the reducing agent. Thus with titanium tetrachloride, bis-π-cyclopentadienyltitanium dichloride, 6.3, is first formed normally, but can react with excess NaC_5H_5 to give bis-π-cyclopentadienyl-σ-cyclopentadienyltitanium, 6.4.

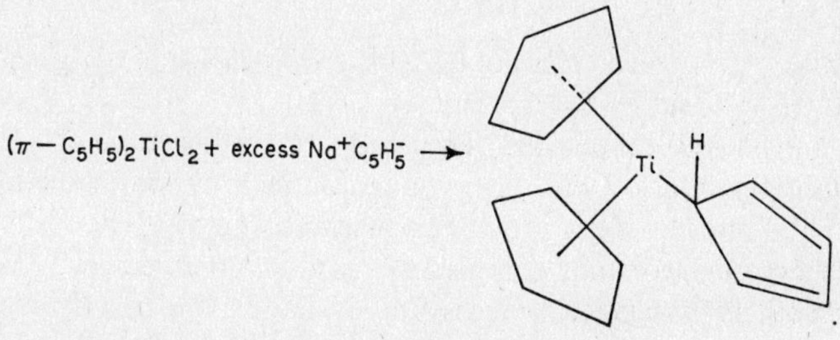

$$(\pi - C_5H_5)_2 TiCl_2 + \text{excess } Na^+C_5H_5^- \longrightarrow$$

6.4

Sometimes a large excess of sodium cyclopentadienide gives π-cyclopentadienyl metal hydrides, e.g.

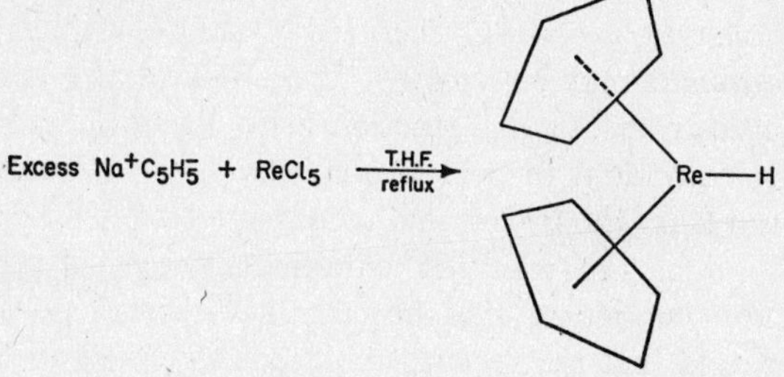

$$\text{Excess } Na^+C_5H_5^- + ReCl_5 \xrightarrow[\text{reflux}]{\text{T.H.F.}} \quad Re\!-\!H$$

Improved yields of bis-π-cyclopentadienylrhenium hydride are obtained when borohydride is present in the reaction mixture.

Treatment of metal carbonyls with alkali metal cyclopentadienides in tetrahydrofuran can give π-cyclopentadienyl metal complexes, e.g.

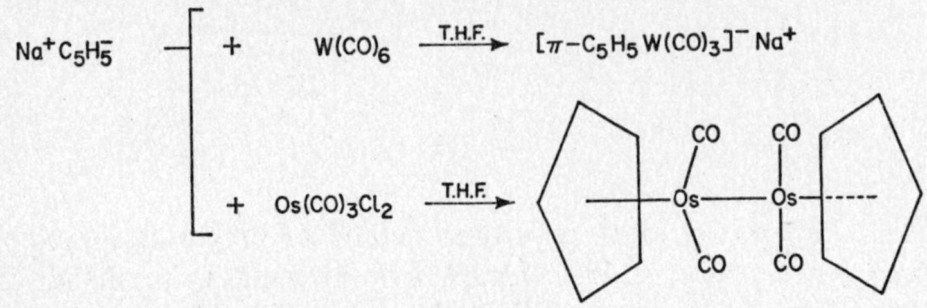

The explosive di-potassium salt of cyclo-octatetraene, $(K^+)_2(C_8H_8)^{2-}$ may similarly be used in the preparation of cyclo-octatetraene metal derivatives, e.g.

$$2(K^+)_2(C_8H_8)^{2-} + MCl_4 \longrightarrow M(C_8H_8)_2 + 4KCl$$
$$(M = Th, U; see Figure 32)$$

Likewise, the green compound π-$C_5H_5TiC_8H_8$ is obtained from $(K^+)_2(C_8H_8)^{2-}$ and $(\pi$-$C_5H_5)_2TiCl_2$.

Hein's 'polyphenyl' complexes. The study of the reaction of anhydrous chromium(III) chloride with phenylmagnesium bromide in diethyl ether by Hein and later by Zeiss provides a fascinating story. The formation of unstable organochromium complexes was first observed in 1903, but it was not until 1919 that Hein successfully isolated a 'polyphenyl chromium complex' from the above reaction. The nature of these organochromium products was not understood, however, until Zeiss and his coworkers (1954) proved that they are sandwich compounds with π-arene ligands such as benzene and biphenyl. Subsequent studies have shown that the formation of these π-arene complexes proceeds *via* chromium σ-aryls, which have been isolated in a few cases. Thus if anhydrous chromium (III) chloride in tetrahydrofuran is treated with phenylmagnesium bromide in an exact 1:3 mole ratio at $-20°C$, deep red crystals of triphenylchromium tristetrahydrofuranate may be isolated. The presence of σ-bonded phenyl groups is demonstrated by its reaction with mercuric chloride, when phenylmercuric chloride is formed in quantitative yield. Tetrahydrofuran may be removed from the tristetrahydrofuranate by washing with diethyl ether. Hydrolysis of the resulting paramagnetic black powder yields π-arene chromium complexes. It is thought that the black powder contains radical species, and that the reaction proceeds as shown in Figure 33.

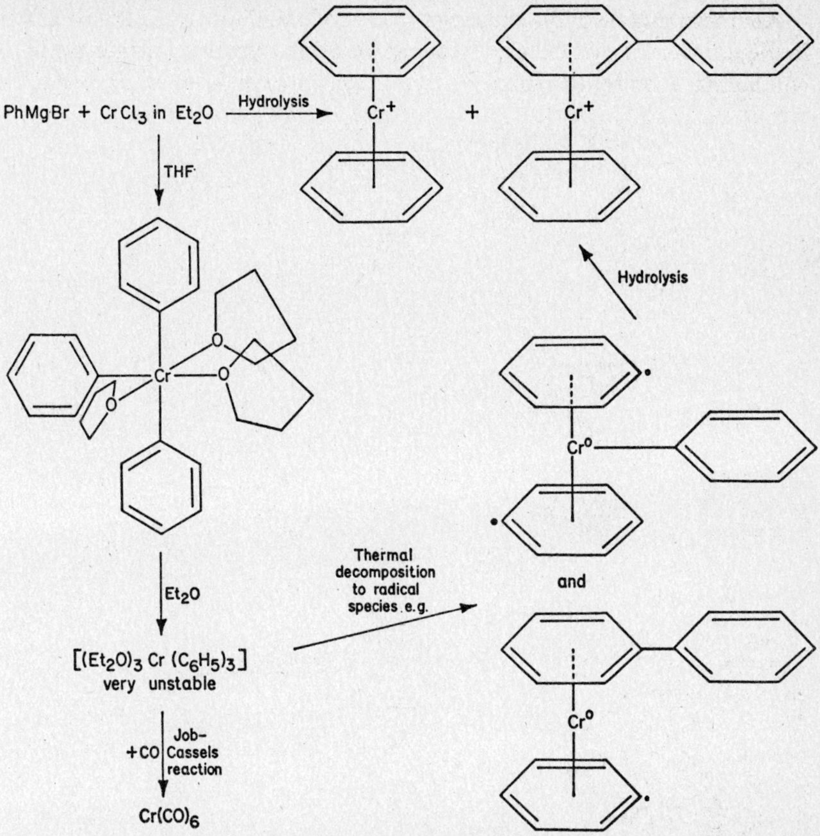

Figure 33. Possible mode of formation of π-arene chromium compounds from σ-bonded phenylchromium derivatives.

Substitution using organometallic derivatives of the transition elements; ligand transfer

A fairly general method for the preparation of tetraphenylcyclobutadiene complexes is provided by ligand transfer reactions. Some of these, using $[\pi\text{-}C_4Ph_4PdBr_2]_2$ as the starting material, are shown in Figure 34. Transfer of the $\pi\text{-}C_5H_5$ ligand may also occur, e.g. nickelocene and iron pentacarbonyl yields 6.5, a 'mixed' metal complex:

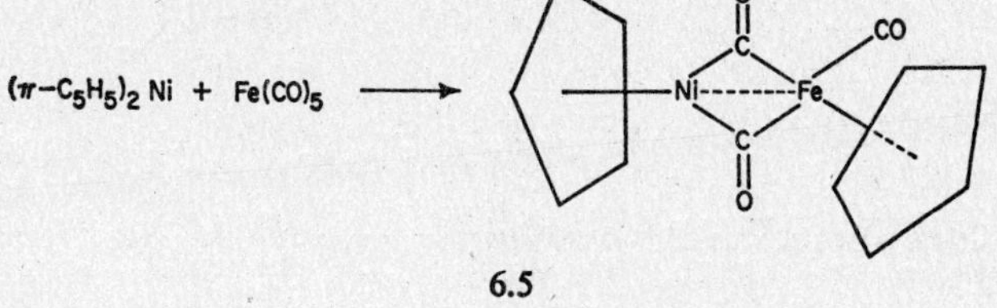

6.5

Another example of the exchange of a π-cyclopentadienyl ligand is shown in Figure 34. Possibly these exchange reactions involve an intermediate containing a bridging π-bonded ligand system, i.e. $M-\pi$-C_nH_n-π-M'.

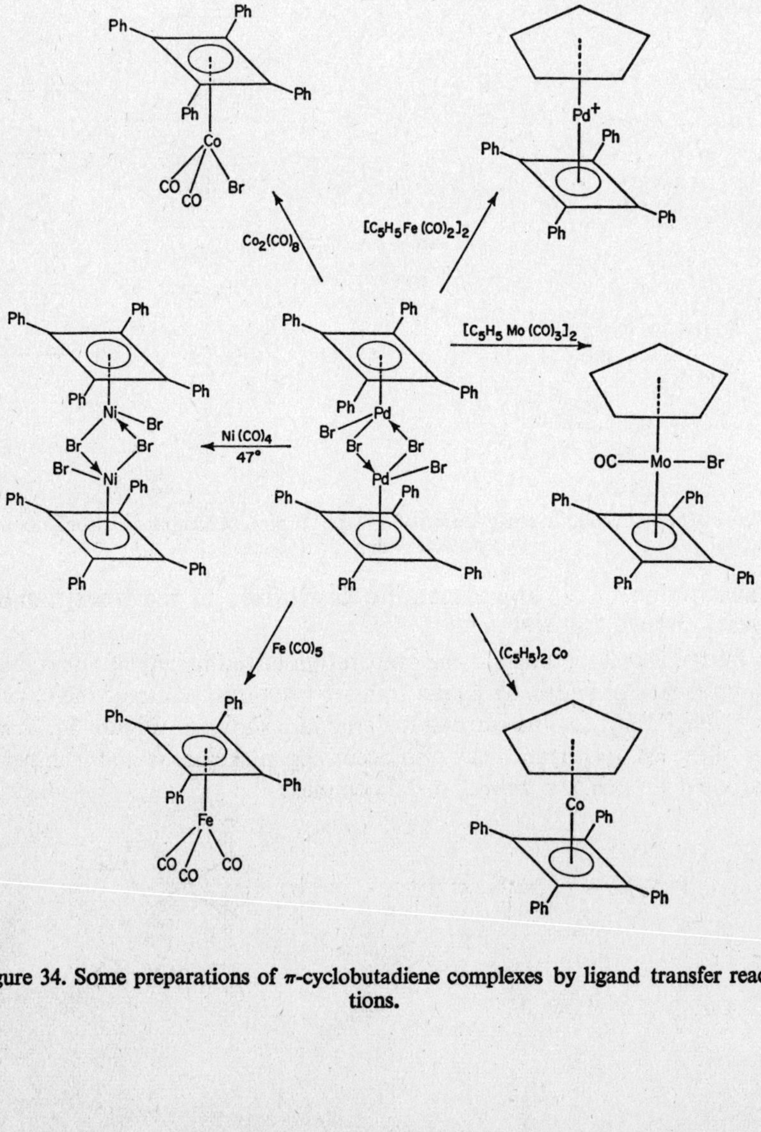

Figure 34. Some preparations of π-cyclobutadiene complexes by ligand transfer reactions.

From complex transition metal anions and halides

Complexes which contain a transition metal in the anion may react with alkyl halides to yield σ-alkyl metal derivatives. Such anions are formed commonly by reduction of metal carbonyls or cyclopentadienyl metal carbonyls with sodium amalgam in tetrahydrofuran:

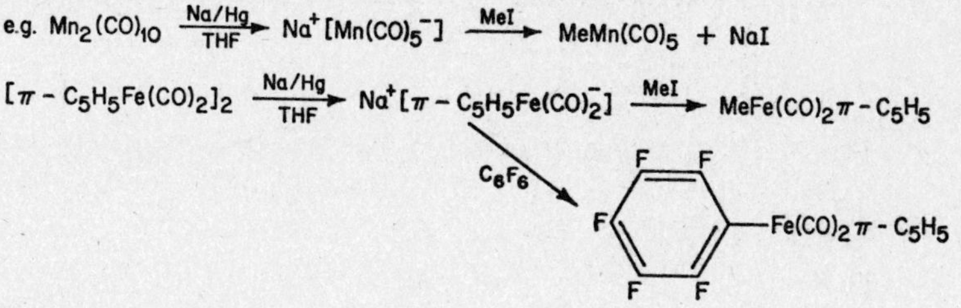

e.g. $Mn_2(CO)_{10}$ $\xrightarrow[THF]{Na/Hg}$ $Na^+[Mn(CO)_5^-]$ $\xrightarrow{MeI}$ $MeMn(CO)_5$ + NaI
$[\pi-C_5H_5Fe(CO)_2]_2$ $\xrightarrow[THF]{Na/Hg}$ $Na^+[\pi-C_5H_5Fe(CO)_2^-]$ $\xrightarrow{MeI}$ $MeFe(CO)_2\pi-C_5H_5$
C_6F_6
F
$Fe(CO)_2\pi-C_5H_5$

With allyl halides a σ-allyl derivative is first formed, which may be converted, by loss of one carbonyl or cyanide group, into π-allyl derivatives, either by heating or by ultraviolet irradiation:

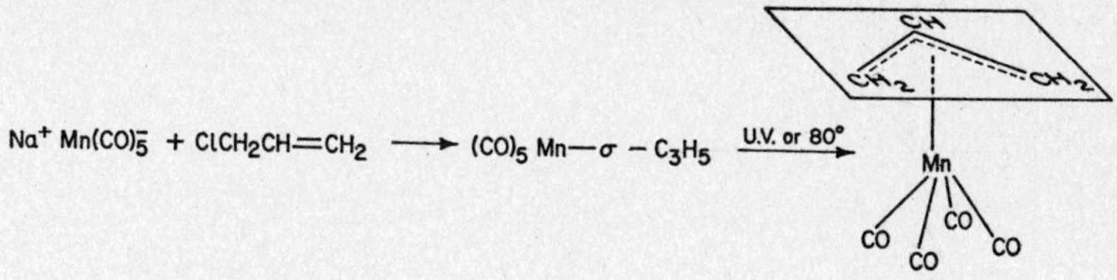

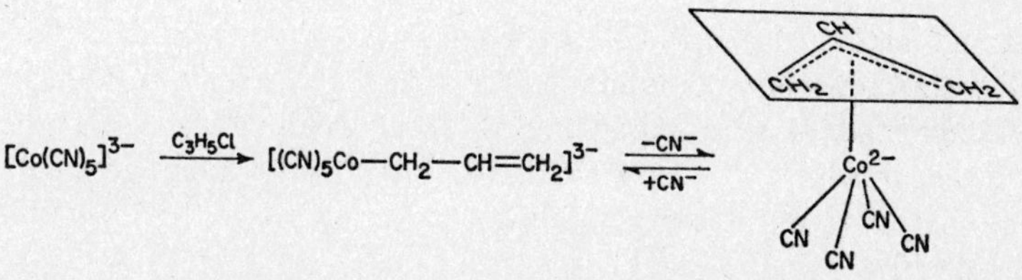

$Na^+ Mn(CO)_5^-$ + $ClCH_2CH{=}CH_2$ → $(CO)_5 Mn—\sigma-C_3H_5$
U.V. or 80°
CH
CH_2
Mn
CO

$[Co(CN)_5]^{3-}$ $\xrightarrow{C_3H_5Cl}$ $[(CN)_5Co—CH_2—CH{=}CH_2]^{3-}$
$-CN^-$
$+CN^-$
CH
CH_2
Co^{2-}
CN

Similarly benzyl chloride reacts with $Na^+[\pi-C_5H_5Mo(CO)_3]^-$ to form the expected σ-benzyl derivative $\pi-C_5H_5Mo(CO)_3\sigma-CH_2Ph$. On irradiation, carbon monoxide is evolved, and a red compound is formed, which has been shown to be the first example of a complex containing the 3-electron π-benzyl system 6.6:

Mo
CH_2
CO

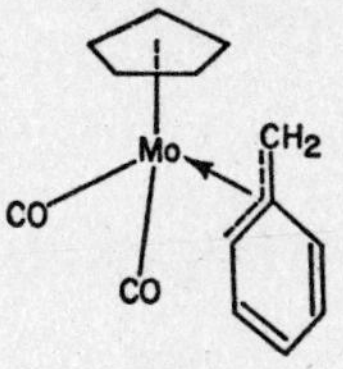

6.6

Unexpected products

On account of the catalytic properties of transition metals unexpected products are frequently obtained in reactions involving their organometallic compounds. For example the reaction of cyclohexene with palladium chloride probably gives initially the expected olefin complex, which is unstable, however, and loses a hydrogen atom to yield a π-enyl complex.

Again, iron pentacarbonyl isomerizes *cis*-substituted conjugated dienes giving products derived from the *trans*-isomer. This may occur since the repulsive interaction between substituents and the metal would be less in this configuration, e.g.

Oligomerization of olefins or acetylenes can also occur in their reactions with transition metal compounds. Treatment of ruthenium trichloride in 2-methoxyethanol with butadiene at 90°C yields the complex 6.7 the structure of which has been determined by X-ray diffraction. (Related reactions involving bis-π-allyl nickel and butadiene are discussed in Chapter 10.)

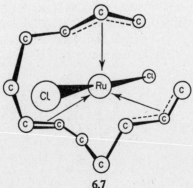

6.7

Acetylenes especially are prone to give unusual reactions, many of which involve oligomerization of the ligand. Examples of some types of products which can be formed in their reactions with compounds of transition metals are given in Table XVIII (Chapter 9).

Most unexpected is the isolation in low (0·5%) yield of a black complex $Fe_5C(CO)_{15}$ from the reaction of $Fe_3(CO)_{12}$ with 1-pentyne. The X-ray crystal structure shows a formally pentaco-ordinate carbon atom located slightly below the plane of the four iron atoms and approximately equidistant from all five iron atoms, 6.8.

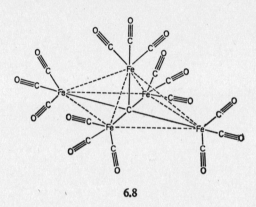

6.8

It is clear therefore that it is very risky to assume, without independent evidence, that when a hydrocarbon ligand reacts with a transition metal its basic structure remains unchanged during the reaction.

BIBLIOGRAPHY

J. M. Birmingham, 'Advances in Organometallic Chemistry, Volume 2, edited by F. G. A. Stone and R. West (Academic Press, New York, 1964), p. 365. Synthesis of π-cyclopentadienyl compounds.

R. E. Dodd and P. E. Robinson, 'Experimental Inorganic Chemistry' (Elsevier, Amsterdam, 1954). A general account of experimental techniques.

'Inorganic Syntheses' (McGraw-Hill, New York). A series which contains experimental details for preparation of inorganic compounds, including some organometallic derivatives.

W. L. Jolly, 'Synthetic Inorganic Chemistry' (Prentice-Hall, London, 1960). General account of experimental techniques.

R. B. King, 'Advances in Organometallic Chemistry, Volume 2, edited by F. G. A. Stone and R. West (Academic Press, New York, 1964), p. 157. Reactions of carbonylate anions.

R. B. King, 'Organometallic Syntheses', Volume I (Academic Press, New York, 1965). Transition Metal Compounds. Detailed preparative procedures.

R. L. Pruett, in 'Preparative Inorganic Reactions', Volume 2, edited by W. L. Jolly (Interscience, New York, 1965), p. 187. Preparation of cyclopentadienyl and arene metal complexes.

CHAPTER SEVEN

Reactions and structures of organometallic compounds of the transition elements

Introduction

In this chapter some of the general reactions and the structures of organo-metallic compounds of the transition elements will be discussed, using the classification of ligands which has been introduced earlier. In this way the most important types of ligand will be exemplified, with emphasis on the structures of their complexes, and on the conversion of one type of ligand into another.

One-electron ligands

As was mentioned in Chapter 1, metal-carbon σ-bonds (e.g. M—Me, M—Et, M—Ph) are formed by most transition elements. Compounds containing these bonds are often isolable at room temperature only when the metal is in particular electronic environments, especially the 18-electron arrangement (see pp 150–157).

The following main classes of complexes containing 1-electron ligands may be recognized:

I (a) Hydrocarbon alkyl and aryl complexes, $M—R_h$.
   (b) Hydrocarbon acyl complexes, $M—COR_h$.
   (c) σ-Alkenyl complexes, $M—CR^1=CR^2R^3$,
      e.g. $\pi\text{-}C_5H_5Fe(CO)_2CH=CH_2$.
   (d) σ-Cyclopentadienyl complexes, e.g. $\pi\text{-}C_5H_5Fe(CO)_2\text{-}\sigma\text{-}C_5H_5$.

II Fluorocarbon alkyl and aryl complexes, $M—R_f$ (e.g. $M—CF_3$, $M—C_6F_5$) and the acyl complexes $M—COR_f$.

III Acetylide or alkynyl complexes, $M—C\equiv CR$.

The bonding of one-electron ligands to transition metals

The possible combinations of metal and carbon orbitals in the formation of a M—C system are shown in Figure 35. It can be seen that contributions to the bonding of both σ- and π-type symmetry can arise. This is emphasized in Table XIV.

Table XIV. *Possible combinations of metal and carbon orbitals in the symmetry group $C_{\infty v}$.*

Symmetry	Metal orbitals	Ligand orbitals	Bond type
A_1	$s, p_z, d_z{}^2$	s, p_z	σ
E_1	p_x, p_y, d_{xz}, d_{yz}	p_x, p_y	π
E_2	$d_{x^2-y^2}, d_{xy}$	—	—

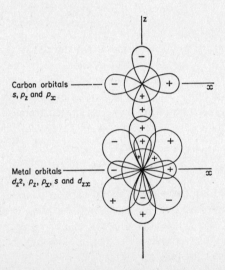

Figure 35. Representation of the interaction between metal and carbon valence orbitals in the *zx*-plane. The relative sizes of the orbitals are unknown. The negative lobe of the $d_z{}^2$ orbital is omitted.

In metal-alkyl (R_h) bonds there is probably little contribution by π-bonding, as the p_x and p_y orbitals of the α-carbon atom, which are of the correct symmetry for π-overlap with metal orbitals are already fully involved in bonding with carbon or hydrogen atoms which are also attached to this atom.

As with all covalent bonds, the metal-carbon bond strength would be expected to increase with increasing electronegativity of the alkyl group;

the greater the electronegativity difference between the alkyl group and the metal group, the larger will be the ionic contribution to the bond strength. Also, strongly polar bonds would be expected to show a reduced tendency towards homolytic dissociation. There are few data available for simple hydrocarbon alkyls with which to test this prediction, but it is observed qualitatively that metal perfluoroalkyl complexes, $M—R_f$, are more stable thermally than the corresponding $M—R_h$ compounds. For example $F_3C—Co(CO)_4$ can be distilled without decomposition at ·91°C, whereas $H_3C—Co(CO)_4$ is stable only at low temperatures (ca. $-30°C$). First the rather high electronegativity of the R_f group (x_{CF_3} = ca. 3·4 cf x_{CH_3} = ca. 2·0) would result in a greater ionic contribution to the bonding than in hydrocarbon alkyls. Secondly the higher positive charge on the metal would cause contraction of the metal orbitals, leading to reduced non-bonding interactions, and to better overlap between metal bonding d-orbitals and suitable ligand orbitals. There is also evidence from infrared spectra and bond lengths which suggests that π-bonding may contribute to the $M—C$ bond in fluorocarbon derivatives, $M—R_f$.

In metal aryls there is the possibility of donation of electrons from the filled $p\pi$-orbitals of the ligand to empty metal orbitals, and also transfer of metal electrons to the empty $p\pi$-antibonding orbitals of the aryl ligand. Commonly transition metal aryls are often more stable thermally than their alkyl analogues. For example the alkyl derivatives $(R'_3P)_2NiR_2$ are very unstable and attempts to isolate them have been unsuccessful. It is possible, however, to obtain a crystalline complex ($R = Ph$), but not in a pure state, as it readily decomposes. ortho-Substituted phenyls (e.g. Ar = mesityl) form surprisingly thermally stable complexes of stoicheiometry $(R_3P)_2NiArX$ and $(R_3P)_2NiAr_2$. These trans complexes are coloured yellow-brown and the majority melt in the range 100–150°C. With few exceptions, they are stable in boiling ethanol and benzene and appear to be stable indefinitely in air in the solid state.

The thermal stability of these ortho-substituted aryls is attributed to both electronic and steric factors. Thus it is argued that the methyl groups hinder attack along the z-axis and also hold the aromatic rings in the zx-plane so that the ring π-orbitals may interact with the metal d_{xy} orbitals (see Figure 36.)

Alkynyl complexes, which contain the grouping $M—C{\equiv}CR$, are often more stable than the corresponding alkyls and aryls. An example is the nickel complex $(Et_3P)_2Ni(C{\equiv}CPh)_2$, which melts at 149–151°C without decomposition. Alkynyl complexes are formally similar to metal cyanides and metal carbonyls. The species $\bar{C}{\equiv}CR$, CN^- and CO are all isoelectronic and presumably bond to a transition metal in a similar manner. All three ligands would be expected to possess both σ-donor and π-acceptor

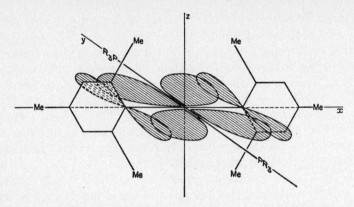

Figure 36. 'Ortho-effect'. The ortho-methyl groups of the mesityl ligands are held in the xz plane by the tertiary phosphine ligands. Thus the d_{xy} orbital must interact with the mesityl $p\pi$-orbitals.

properties, and a conventional representation of the bonding in the systems M—C≡X, where X = O, N, or CR is shown in Figure 37.

Evidence for π-donation from metal orbitals into the antibonding molecular orbitals of the alkynyl group has been obtained from studies of the infrared spectra of such complexes. For the platinum compounds $(Et_3P)_2Pt(C≡CR)_2$ the C≡C stretching frequencies occur $ca.$ 2100 cm^{-1} which is about 150 cm^{-1} lower than those found in disubstituted acetylenic hydrocarbons (2260 – 2190 cm^{-1}), suggesting a lowering of the C≡C bond order. The C≡C distance in the compound $(Et_3P)_2Ni(C≡CPh)_2$ however, is very close (1·18 ± 0·02 Å) to that found in the free acetylene PhC≡CPh (1·19 ± 0·03 Å).

Two other ligands which are formally similar to CO, CN$^-$ and C≡CR are the nitrogen molecule N≡N and the thiocarbonyl group CS. Some

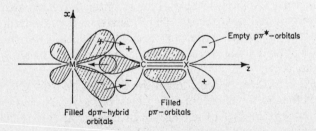

Figure 37. A conventional representation of the bonding in M—C≡X systems where X = O, N or CR. Only the π-bonding in the xz-plane is shown. A similar interaction is present in the yz-plane. This bond description may be compared to that of the M-ethylene bond (See Figure 29).

complexes of molecular nitrogen are discussed in Chapter 10. An example of a complex containing the thiocarbonyl ligand is 7.1.

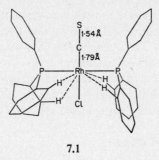

7.1

Metal 'carbenes'

The structure of the compound W(CO)₅-CPhOMe, determined by X-ray analysis, is shown in Figure 38.

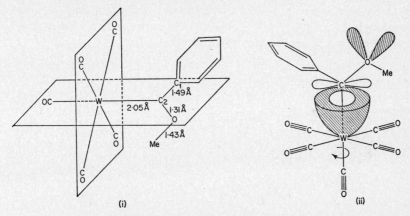

(i)   (ii)

Figure 38. (i) Structure of PhMeOCW(CO)₅, showing that the OC₁C₂ plane bisects the cis-axes, through the metal-carbonyl ligands.
(ii) The 'cylindrical' π-bonding orbital is represented. The x, y and z axes lie along the W—C bonds.

This compound illustrates the stabilization by complex formation of a species which is unstable in the free state.

The W—C bond of the W—CPh(OMe) system (2·05 Å) is rather longer than those of the W—CO systems (~1·89 Å). This W—C₂ bond is probably quite polar; in the extreme case this might be represented as an 'ylide' system, $(OC)_5W^+—\bar{C}(OMe)Ph$. In addition there is likely to be considerable double bonding between the tungsten

d-orbitals and the C_2 p-orbitals which are normal to the C_1C_2O plane. The plane of the C_1C_2O group lies between the x and the y axes at about 45° to each. This structure is the one to be expected, since in the group $M(CO)_5$ all the metal orbitals are symmetrically distributed about the z-axis and hence both the d_{xz} and d_{yz} and the p_x and p_y metal orbitals form degenerate pairs. These metal orbitals therefore form a symmetrical, cylindrical π-bonding orbital about the z-axis (see Figure 38(ii)) and thus in this situation a M=C double bond has, apart from steric effects, free rotation about the M—C axis. It follows that, as found, the OC_1C_2 plane would be expected to take up the least sterically hindered position.

Properties of the M—C bond

It is very often found that the metal-carbon bond is cleaved by halogens and less often by acids such as the hydrogen halides: e.g.

$$cis\text{-}(Et_3P)_2PtBr_2 \xleftarrow{\quad Br_2 \quad} cis\text{-}(Et_3P)_2PtMe_2 \xrightarrow{\quad HCl \quad} cis\text{-}(Et_3P)_2PtMeCl$$

Where a β-hydrogen atom is present in the alkyl group (e.g. ethyl, n-propyl, isopropyl), treatment with a hydride ion abstractor such as the triphenyl-methyl cation can lead to ethylenic cations, e.g.

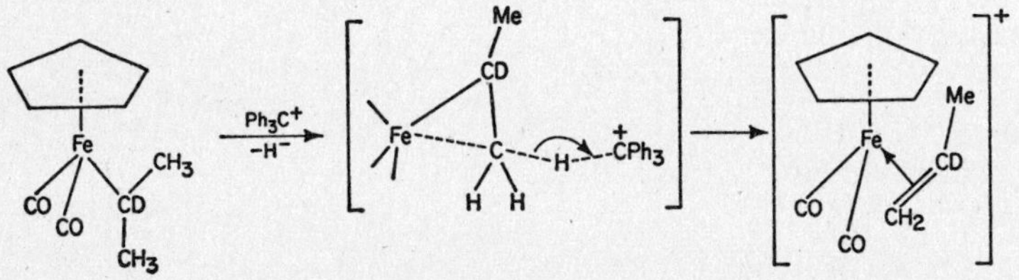

7.3   7.2

The formation of the 2-deuteropropene cation, 7.2, from the 2-isopropyl complex, 7.3, as shown above, confirms that the hydride is abstracted from a β-carbon atom. Similarly, the thermal decomposition of the ethyl complex *trans*-$(Et_3P)_2PtCl(CH_2CH_3)$ to the hydride *trans*-$(Et_3P)_2PtHCl$ and ethylene may proceed via internal abstraction of a hydrogen from the β-carbon, viz.

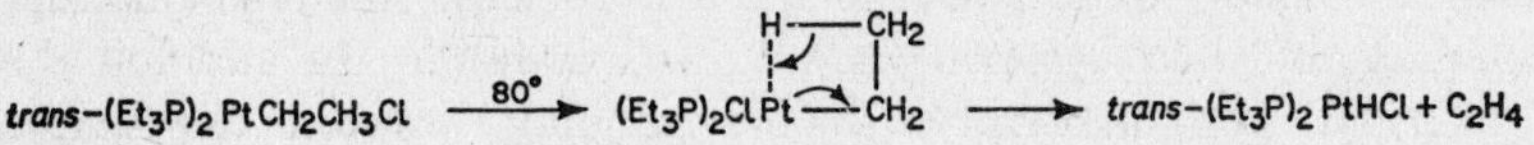

Where the organic ligand has an unsaturated β-carbon atom, e.g. metal σ-allyl complexes, protonation can lead to π-olefinic cations, e.g.

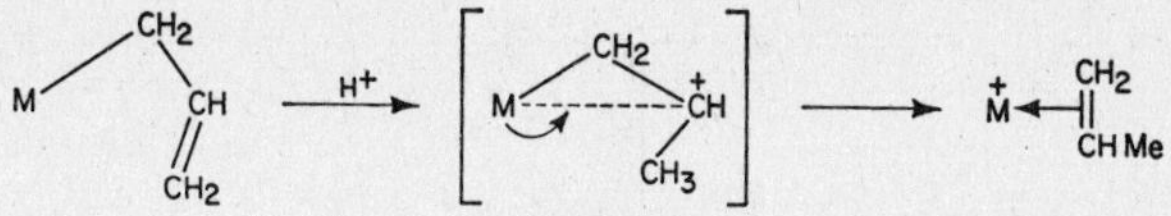

Postulated
mechanism

where, $M = Mn(CO)_5$; π-$C_5H_5Fe(CO)_2$.

It may be envisaged that both the hydride abstraction and the protonation reactions proceed *via* the formation of a β-carbonium ion intermediate which is 'stabilized' by formation of an ethylene-metal bond and partial oxidation of the metal, e.g.

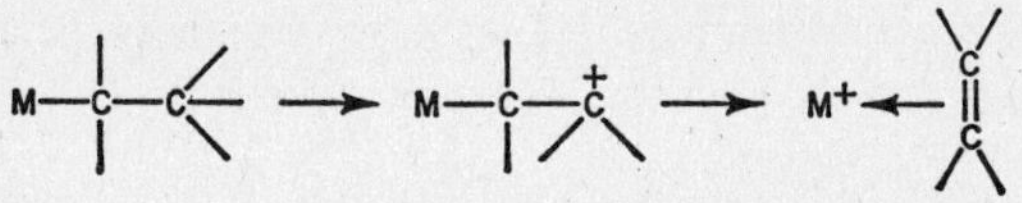

It appears that the stabilization of β-carbonium ions by transfer of some positive charge to the metal and formation of a π-bond between the metal and organic ligand is an important 'group property' of complexed transition metals.

The above examples illustrate the conversion of a formal 1-electron ligand (e.g. —CH_2CH_3) to a 2-electron ligand (e.g. $H_2C{=}CH_2$). Several related examples of the conversion of an n-electron ligand in a complex either to an $(n-1)$ or to an $(n+1)$-electron ligand will be described later in this chapter.

Insertion reactions

Another important property of transition metal alkyl and aryl bonds is their ability to undergo insertion reactions, e.g.

$$M{-}R_h + X \longrightarrow M{-}X{-}R_h,$$

where $X = CO$, SO_2, $F_2C{=}CF_2$.

In the carbonylation reaction, in which a metal alkyl or aryl is converted into the corresponding acyl derivative, isotopic tracer studies have shown that it is a co-ordinated CO group which inserts into the metal-carbon bond by an intramolecular shift, and not a CO molecule from the gas phase:

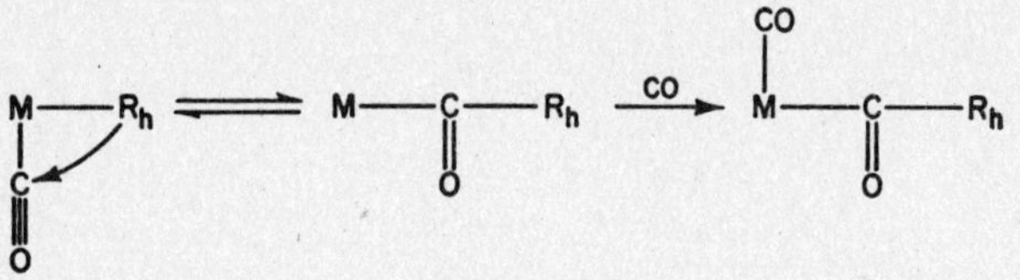

Carbonylation reactions, especially their application in the industrially important OXO process, are discussed further in Chapter 10.

Two-electron ligands

The bonding and structures of π-olefin complexes of the transition elements have already been discussed in some detail in Chapter 5, and their preparation in Chapter 6. We have shown above (p 189) that 2-electron ligands can be formed from and can themselves be converted into 1-electron ligands. The following example shows how one may proceed from a coordinated olefin to an enyl or 3-electron system:

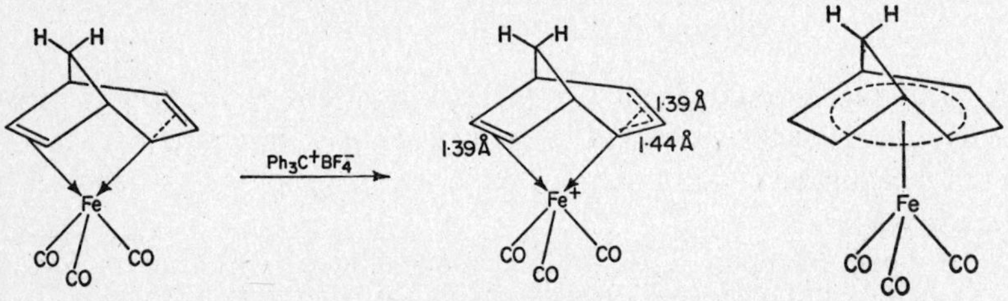

The crystal structure of the complex, 7.4, has been determined and shows that the ligand may be regarded either as a combination of 2- and 3-electron, 7.4, ligands or as a 5-electron homo-aromatic ligand, 7.5. Examples of $n \times 2$-electron ligands are given in Figure 32, p 166.

Three-electron ligands; π-allyl and π-enyl complexes

Aspects of the chemistry of π-enyl complexes treated elsewhere in this book are, bonding, p 164; preparation, p 174; occurrence in catalytic reactions, p 239.

    A typical π-enyl complex, whose structure has been determined by X-ray diffraction is bis-π-2-methylallylnickel, 7.6. The planes of the two allyl groups are parallel, the nickel atom being equidistant from these planes in a sandwich structure.

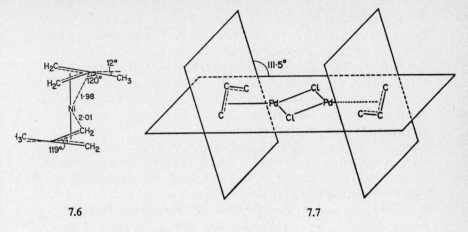

7.6   7.7

The dimer $(\pi\text{-}C_3H_5PdCl)_2$ has the configuration 7.7. The plane of the three allyl carbon atoms intersects the plane formed by the $(PdCl)_2$ system at an angle of about 111·5°, with the central carbon tipped away from the palladium. The $(PdCl)_2$ plane is nearer the two terminal carbons than the central carbons of the allyl groups. At -140°C the Pd to central carbon distance is the same as that of the equivalent Pd to terminal carbon distances (av 2·11 Å). At room temperature the Pd to central carbon distance may be marginally longer but not by more than 0·1 Å. The apparent increase of distance may arise from errors caused by thermal motions at higher temperatures. The C—C distances of the π-allyl groups are the same (av 1·36 Å) and the Pd—Cl distances (av 2·41 Å) are similar to those found in other complexes for Pd—Cl *trans* to olefin ligands. A low temperature study locates approximately the hydrogen atoms of the allyl groups, and, within experimental error, they are found to lie in the plane of the allyl carbons.

π-1-Methylallyl cobalt tricarbonyl is obtained as a mixture of geometrical isomers, 7.8, and 7.9, from the reaction between $HCo(CO)_4$ and butadiene. The *syn*- isomer is more stable and may be obtained from the *anti*- isomer by equilibration at 80°.

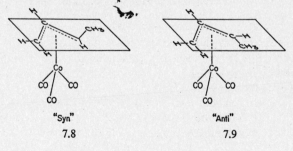

"Syn"   "Anti"
7.8   7.9

Dynamic allyls

The proton magnetic resonance spectra of allylic transition metal complexes fall essentially into three classes. The σ-bonded allyls give the spectra to be expected from their formulation $M\text{—}CH_2CH\text{=}CH_2$. Most π-allyl groups, e.g. in $\pi\text{-}C_3H_5Mn(CO)_4$, show spectra which are that of an A_2B_2X system; a typical spectrum of this type is shown in Figure 39 (i). The third class of spectra suggest that the allyl groups have equivalent terminal hydrogens, that is, spectra typical of an A_4X system are found, (Figure 39 (ii)). Some allyl complexes show the π-allyl spectra A_2B_2X at lower temperatures and the A_4X spectra at higher temperatures, e.g. $(allyl)_3Rh$ and $(allyl)_2M$ (M = Ni, Pd or Pt). It is suggested that at the higher temperatures there is rapid movement of the allyl groups about the metal atom, so that the environment of the terminal protons becomes averaged within the relatively long time scale of the nuclear magnetic resonance experiment.

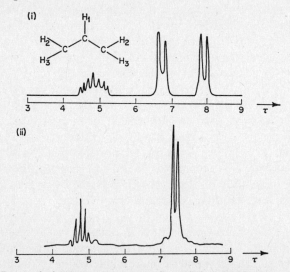

Figure 39. (i) A typical ^1H n.m.r. spectrum of a π-allyl, A_2M_2X group.
(ii) ^1H n.m.r. spectrum of $Zr(C_3H_5)_4$ at $-20°C$; an A_4X system.

Other allyl complexes, e.g. the π-allyl palladium chloride dimer show the A_4X spectra only in the presence of ligands such as R_3P, R_3As or dimethylsulphoxide. In this case it is concluded that the equilibration of the terminal hydrogens in the n.m.r. experiment is due to rapid exchange processes, e.g.

$$\tfrac{1}{2}(\pi\text{-allyl PdCl})_2 + L \rightleftharpoons \pi\text{-allyl PdClL}$$
$$\pi\text{-allyl PdClL} + L \rightleftharpoons \sigma\text{-allyl PdClL}_2$$
$$\pi\text{-allyl PdClL} + L' \rightleftharpoons \pi\text{-allyl PdClL'} + L \ (L = L')$$

In other words, it is *not* necessary to postulate in these systems any allyl-metal complexes which contain other than σ-allyl or π-allyl-metal bonds.

When treated with an equimolar quantity of triphenylphosphine the π-methylallylpalladium chloride dimer gives the complex 7.10. Proton magnetic resonance studies suggest that the allyl is bound unsymmetrically using both a σ- and an ethylenic, 2-electron π-bond rather than by a symmetrical π-allyl system. X-Ray diffraction has confirmed that the C—C distances of the allyl group are significantly different. Again we are faced with a π-bonding system which cannot be represented adequately by a simple valence bond diagram. The valence bond way to indicate the inequality in the C—C lengths is to draw the σ-, π-system shown in 7.11. Compare allyl-mercuric iodide, Chapter 3, p 67.

This representation is probably not accurate since a considerable degree of π-bonding is still expected across the longer, formally single, C—C bond.

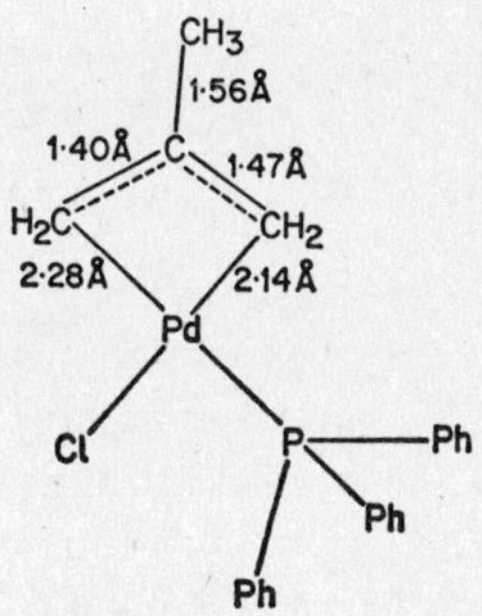

7.10

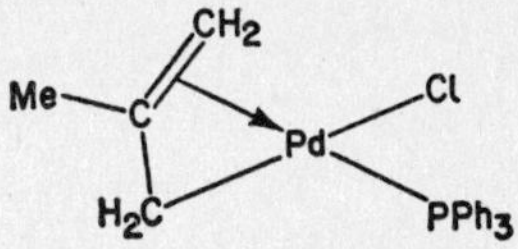

7.11

Four-electron (diene) ligands

Cyclobutadiene complexes

In 1956 Longuet-Higgins and Orgel predicted that the unstable cyclobutadiene system would be stabilized by complex formation with a transition metal. Three years later Criegee isolated the first π-cyclobutadiene complex, 7.12, as a red-violet powder, soluble in methylene chloride and in water. The C_4Me_4 group is nearly planar, and the annular C—C bonds are all equal in length.

The unsubstituted π-cyclobutadiene complex, 7.13, has since been prepared by a method analogous to that used for the nickel complex, and its structure determined by X-ray diffraction.

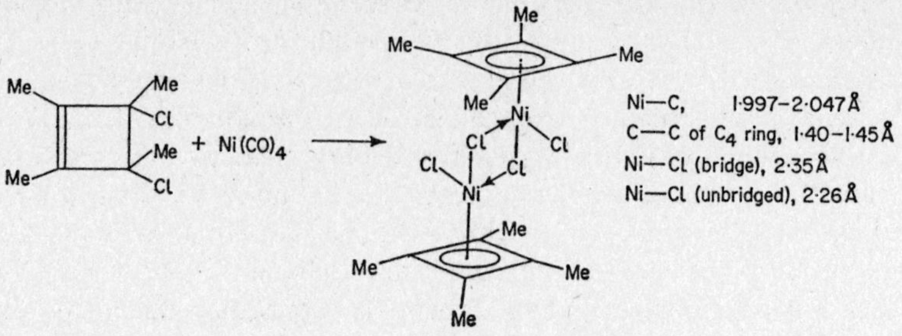

Ni—C,   1·997-2·047Å
C—C of C_4 ring, 1·40-1·45Å
Ni—Cl (bridge), 2·35Å
Ni—Cl (unbridged), 2·26Å

7.12

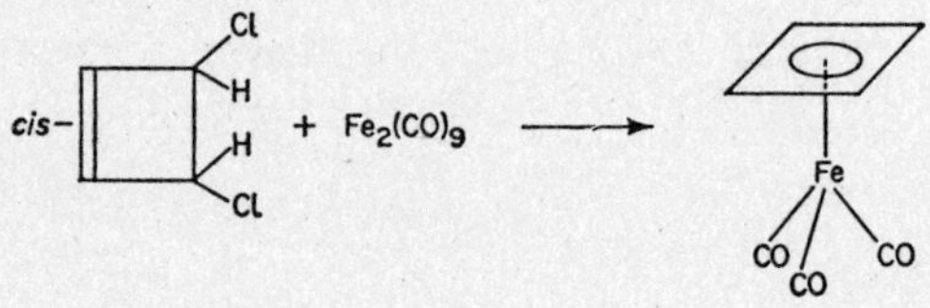

7.13

Oxidation of $C_4H_4Fe(CO)_3$ with ceric ion releases the unstable ligand cyclobutadiene, which may be characterized by *in situ* reactions, for example by Diels-Alder addition to $HC\equiv CCOOMe$, which occurs in a stereospecific manner to give a Dewar benzene derivative, **7.14**,

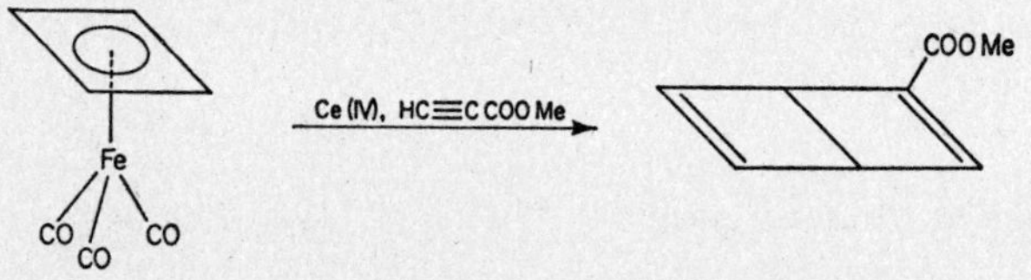

7.14

In fact, some addition reactions of the liberated cyclobutadiene suggest it to be a diene with a rectangular, singlet state structure rather than a delocalized aromatic ring with a square, triplet state structure.

The cyclobutadiene ring in $C_4H_4Fe(CO)_3$ possesses aromatic properties and readily undergoes electrophilic substitution (p 217).

Hydrolysis of the chloromethyl derivative, **7.15**, may involve a π-enyl intermediate, **7.16**.

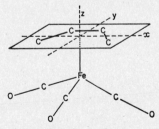

An interesting 4-electron ligand is present in the trimethylenemethane derivative, 7.17.

7.17

Butadiene complexes

The conjugated diolefin, butadiene, may be considered formally as a 4-electron donor. Treatment of $Fe(CO)_5$ with butadiene under pressure yields butadiene iron tricarbonyl, an 18-electron complex, the crystal structure of which is given in Figure 40. The carbon atoms of the butadiene adopt a *cis* configuration and are coplanar. The plane of the four carbon atoms is not quite parallel to that formed by the carbon atoms of the three carbonyl groups. The iron atom lies below the plane of the butadiene ligand and is equidistant from each of the four carbon atoms ($2 \cdot 1 \pm 0 \cdot 04 \text{Å}$). A bonding description of butadiene complexes has been given in Figure 31, Chapter 5.

Complexes of conjugated olefins may react to give complexes containing ligands bound by an essentially delocalized system with either one

Figure 40. The structure of butadiene iron tricarbonyl (after Mills and Robinson).

o

electron more or one less than the parent ligand. Thus 4-electron ligands may be converted by hydride abstraction to 5-electron π-dienyl ligands, while treatment with acids, e.g. HCl, yields π-enyl (3-electron) derivatives.

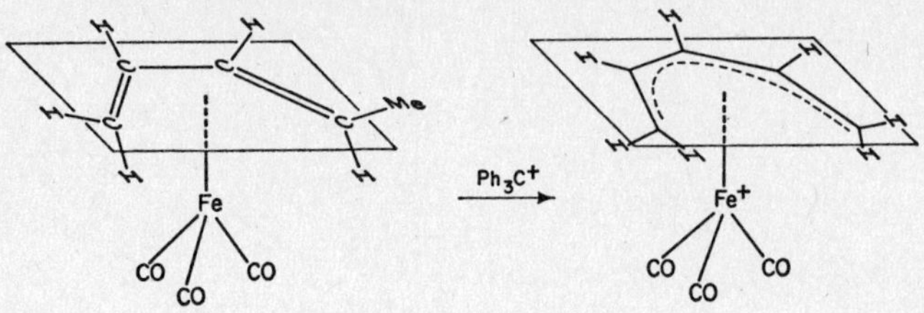

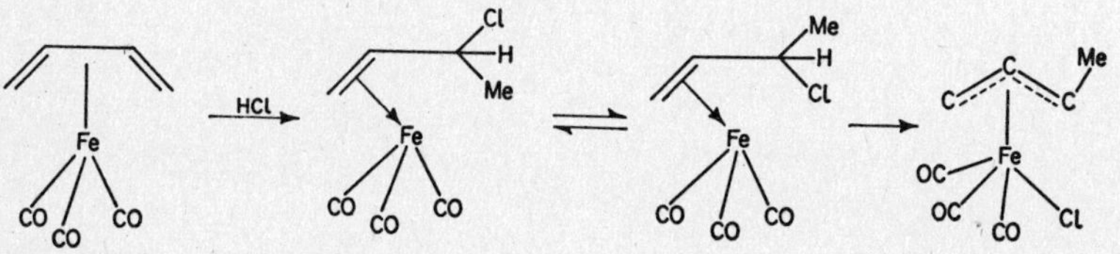

These reactions may be compared with the conversion of mono-ene 2-electron ligands either into 1-electron (alkyl) or 3-electron (π-enyl) systems mentioned above. Understandably, however, unconjugated diene and triene ligands, in which the C=C systems are separated by more than one carbon atom may not react so readily to form delocalized systems.

The structure of π-cyclopentadienyl hexatrifluoromethylbenzene rhodium, 7.18, is of some interest. Normally benzenoid aromatic hydrocarbons

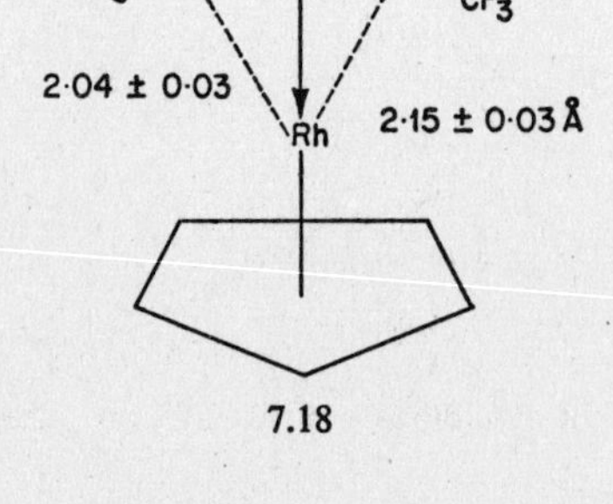

7.18

formally donate 6 electrons to transition metals, the benzene ring remaining planar on co-ordination (see p 208). However, the crystal structure of 7.18 shows that only four carbon atoms of the $(CF_3)_6C_6$ ring are in a plane, the remaining two being bent upwards away from the metal at an angle of $48° \pm 2°$. This structure is in agreement with the 18-electron rule, that is, the hexatrifluoromethylbenzene ligand acts as a conjugated diene in this complex, and donates only four electrons (i.e. Rh = 9; π-C_5H_5 = 5; $(CF_3)_6C_6$ as diene = 4; giving a total of 18 electrons).

Five-electron (dienyl) ligands

As was pointed out in Chapter 5, with reference to the conjugated ligand cyclooctatetraene, and as the structures of the cyclobutenyl nickel complex, 7.15, and of the rhodium complex 7.18 show, it is not necessary for all the annular carbon atoms of a cyclic unsaturated hydrocarbon ligand to be involved in bonding to a metal. Further examples of this are shown in Figure 41, where in complexes derived from 5-, 6- and 7-membered ring systems respectively, only 5 electrons are formally involved in the bonding in each case.

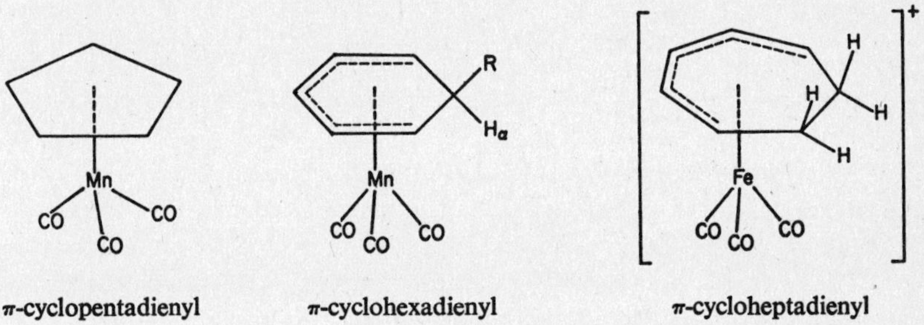

π-cyclopentadienyl   π-cyclohexadienyl   π-cycloheptadienyl

Figure 41. Some 'dienyl' complexes showing the dienyl system in 5-, 6-, and 7-membered cyclic hydrocarbons.

π-Cycloheptadienyl metal complexes may be obtained either by hydride abstraction from a diene-iron complex or by protonation of a suitable triene complex:

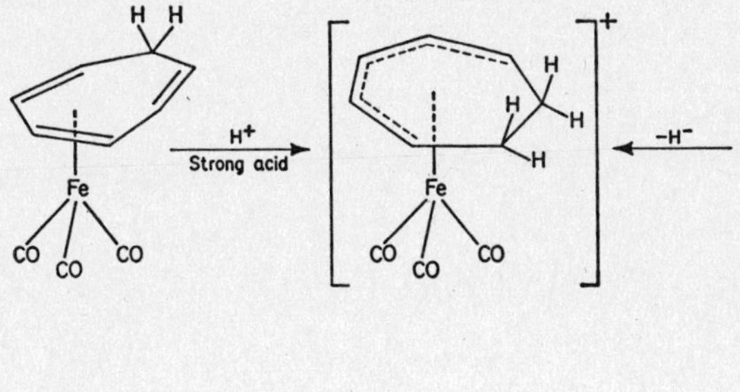

π-Cyclohexadienyl complexes can similarly be obtained by hydride abstraction from dienes, and also result from nucleophilic attack of hydride ion on triene or arene metal cations, e.g.

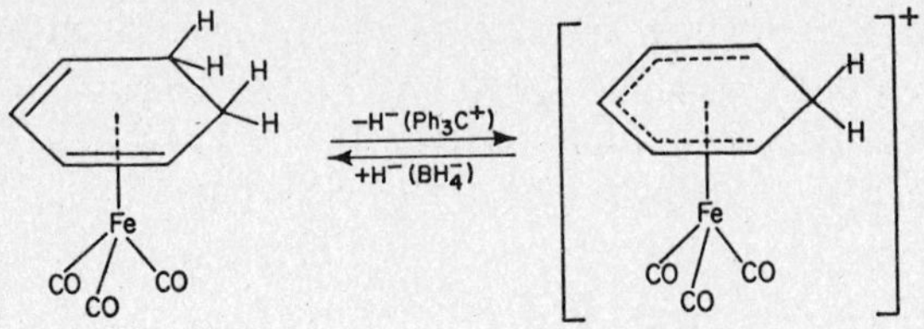

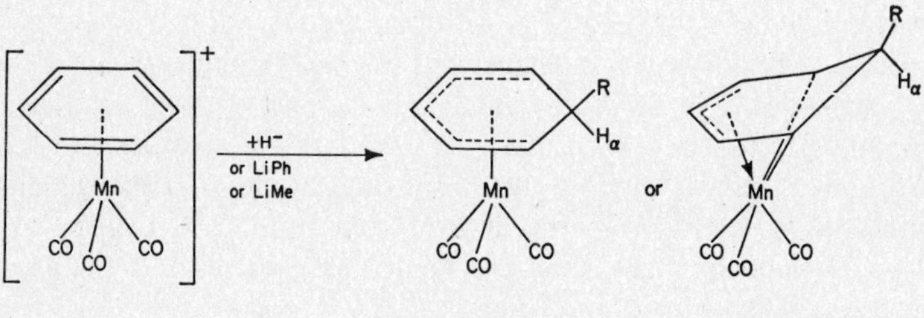

7.20   7.19

The bonding of the π-cyclohexadienyl ligand might be thought of as intermediate between the extreme cases of two σ-bonds and a π-enyl bond, 7.19, and a delocalized C_5 system, 7.20.

An interesting property of some cyclopentadiene- and cyclohexadienyl-metal complexes is the occurence of unusually broad, low frequency bands (2700 cm^{-1}) which arise from a C—H stretch of the methylene carbon. The crystal structure of the rhenium complex, 7.21, which also has an unusually low C—H stretch, shows that the C—hydrogen is in the *exo*-position. It is reasonable to suppose that the unusual C—H stretches in the related complexes also arise from the C—H$_{exo}$ bond. It has been suggested that the C—H$_{exo}$ frequency is lowered as a result of direct interaction between the metal and the methylene carbon.

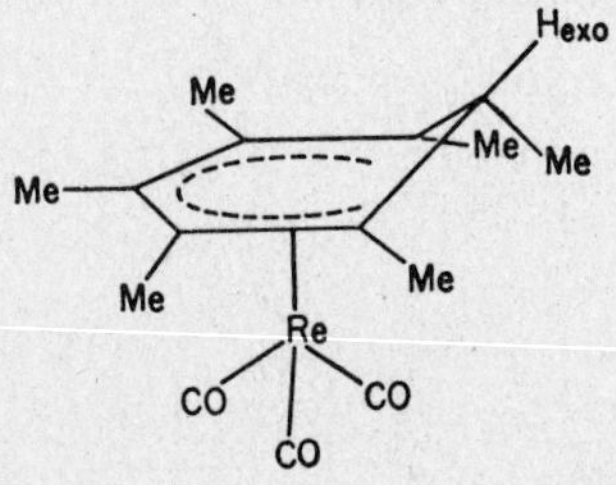

7.21

π-Cyclopentadienyl complexes

The cyclopentadienyl radical $C_5H_5^.$ forms two distinct classes of complex with transition metals. The first and by far the most important of these consists of π-cyclopentadienyl complexes in which the C_5H_5 ring is bonded essentially covalently to the metal, the metal atom lying below the plane of the ring, usually equidistant from all five carbon atoms. In addition, a few σ-cyclopentadienyl complexes are also known, in which the C_5H_5 group is bonded to the metal through one carbon atom only, by a 2-electron covalent bond.

As discussed on p 5, the cyclopentadienyl radical can readily accept an electron forming the cyclopentadienide anion. Both chemical evidence and, especially, magnetic susceptibility data suggest that manganocene is largely ionic and is a cyclopentadienide, $Mn^{2+}(C_5H_5^-)_2$. The rare earth complexes, $(C_5H_5)_3M$, all of which are known, are also thought to be largely ionic in nature.

Complexes containing the covalently bound π-cyclopentadienyl ligand are known for nearly all the *d*-block transition elements. Bis-π-cyclopentadienyl complexes $(π-C_5H_5)_2M$, and cations derived from them (e.g. $[(π-C_5H_5)_2Co]^+$) which contain the $π-C_5H_5$ ligand only, and a very wide variety of derivatives which contain one or two $π-C_5H_5$ groups together with other ligands have been prepared. Some examples are given in Figure 42. The π-cyclopentadienyl-metal bond has particularly high kinetic stability to thermal decomposition, and this accounts for the very diverse range of compounds known containing it.

(*i*) *Bis-π-cyclopentadienyl complexes.* The crystalline bis-π-cyclopentadienyls of the elements of the first transition series, $(π-C_5H_5)_2M$, (M = V, Cr, Fe, Co, Ni) are isomorphous. X-Ray diffraction shows that crystalline ferrocene $(π-C_5H_5)_2Fe$, has the staggered form, and electron diffraction data show that in the vapour phase it has the eclipsed form.

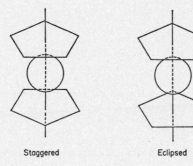

Staggered   Eclipsed

Ruthenocene and osmocene, however, crystallize in the eclipsed form. The C_5H_5 rings in π-cyclopentadienyl complexes are planar and all the C—C bond distances are equal. As will be seen later (p 219) they possess considerable aromatic character, and in favourable cases, e.g. ferrocene, undergo electrophilic substitution reactions such as acetylation.

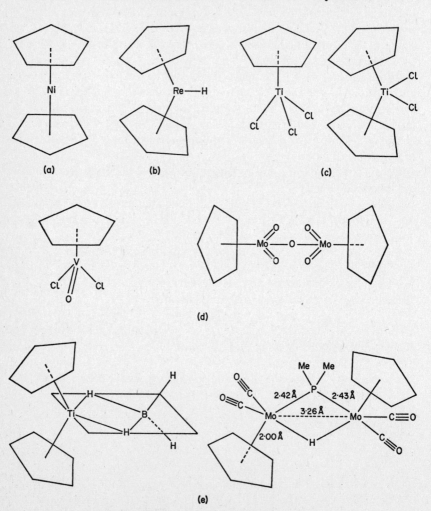

Figure 42. The structures of some π-cyclopentadienyl complexes.
    (a) bis-π-cyclopentadienyl nickel.
    (b) a bent bis-π-cyclopentadienyl hydride.
    (c) π-cyclopentadienyl halides.
    (d) π-cyclopentadienyl metal oxides and oxyhalides.
    (e) π-cyclopentadienyl metal hydrides containing bridging hydrogen atoms.

Table XV. *Magnetic data and electron assignment for bis-π-cyclopentadienyl complexes*

Compound[a]	Electron assignment[b] $(a_{1g})^2(a_{2u})^2$ $(e_{1u})^4(e_{1g})^4$	Number of unpaired spins	Spin only value	Magnetic moment expected	Magnetic moment found
Cp_2Ti^+	$(e_{2g})^1$	1	1.73	>1.73	2.29±0.05
Cp_2V^{2+}	$(e_{2g})^1$	1	1.73	>1.73	1.90±0.05
Cp_2V^+	$(e_{2g})^2$	2	2.83	~2.83	2.86±0.06
Cp_2V	$(e_{2g})^2(a'_{1g})^1$	3	3.87	~3.87	3.84±0.04
Cp_2Cr^+	$(e_{2g})^2(a'_{1g})^1$	3	3.87	~3.87	3.73±0.08
Cp_2Cr	$(e_{2g})^3(a'_{1g})^1$	2	2.83	>2.83	3.20±0.16
Cp_2Fe^+	$(a'_{1g})^2(e_{2g})^3$	1	1.73	>1.73	2.34±0.12
Cp_2Fe	$(a'_{1g})^2(e_{2g})^4$	0	0	0	0
Cp_2Co^+	$(a'_{1g})^2(e_{2g})^4$	0	0	0	0
Cp_2Co	$(a'_{1g})^2(e_{2g})^4$ $(a''_{1g})^1$	1	1.73	~1.73	1.76±0.07
Cp_2Ni^+	$(a'_{1g})^2(e_{2g})^4$ $(e''_{1g})^1$	1	1.73	>1.73	1.82±0.09
Cp_2Ni	$(a'_{1g})^2(e_{2g})^4$ $(e'_{1g})^2$	2	2.83	~2.83	2.86±0.11

[a]$Cp_2 = (\pi\text{-}C_5H_5)_2$,   [b]See Figure 26c.

The sandwich bis-π-cyclopentadienyl complexes are thermally rather stable, and many melt without decomposition at about 173°C. They are stable to hydrolysis, and the C_5H_5 rings resist catalytic hydrogenation. Their stability to oxidation, however, varies greatly with the nature of the metal. At room temperature the 18-electron complex ferrocene is inert to molecular oxygen, whereas chromocene (which possesses a 16-electron configuration) is pyrophoric in air.

In Table XV magnetic data for some metallocenes and metallocenium cations are listed. It can be seen that the experimental magnetic moments are, in many cases, close to those expected using the 'spin-only' formula.

In acid solutions ferrocene is oxidized to dichroic deep blue green (blood-red when concentrated) solutions of the ferricenium ion $[(\pi\text{-}C_5H_5)_2Fe]^+$. Cobaltocene is especially readily oxidized, losing one electron to attain the noble gas configuration. The resulting cobalticenium ion $[(\pi\text{-}C_5H_5)_2Co]^+$ is very resistant to further oxidation; it is, for example, stable in concentrated nitric acid. The cobalticenium cation can be regarded as a large pseudo-alkali metal cation. Nickelocene is also easily oxidized, but the paramagnetic $[(\pi\text{-}C_5H_5)_2Ni]^+$ which is formed (one unpaired electron) is rather unstable and decomposes in solution.

Cobaltocene undergoes several interesting reactions in which the products attain the stable 18-electron configuration. As has been mentioned, this can occur by the loss of the unpaired electron to yield the cobalticenium ion. Sometimes, however, one of the π-cyclopentadienyl groups may be converted into a 4-electron π-cyclopentadiene ligand by addition to the ring:

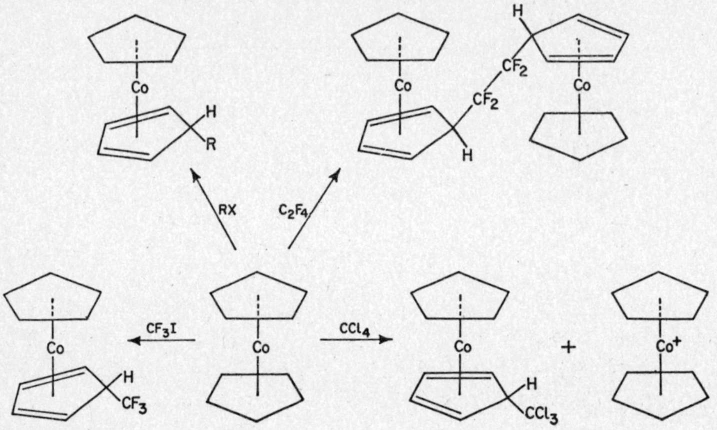

Rhodocene is even more reactive than cobaltocene, and can be isolated only in the absence of solvents, by reduction of the $[(\pi\text{-}C_5H_5)_2Rh]^+$ cation with sodium vapour, followed by vacuum sublimation on to a cold finger. It behaves as a radical, dimerising rapidly at room temperature to give 7.22, in which the rhodium atom has attained the 18-electron configuration.

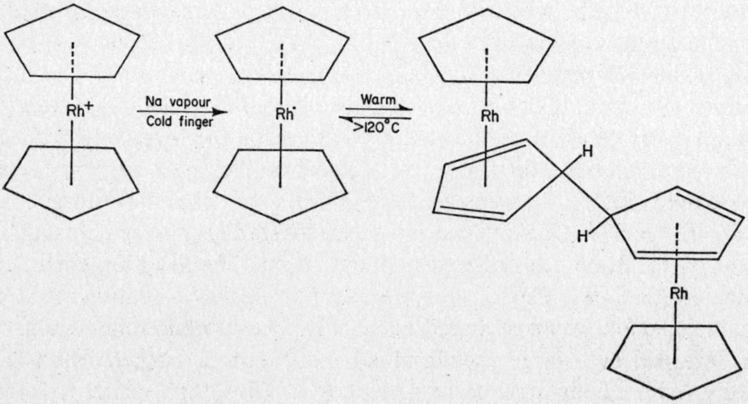

7.22

Similarly nickelocene reacts to give products in which one π-cyclopentadienyl ring is converted to a 3-electron π-enyl ligand.

(ii) π-Cyclopentadienyl metal carbonyls and related compounds

A very wide variety of π-cyclopentadienyl metal carbonyl complexes is known. The structures of some are illustrated in Figure 43. With very few exceptions these compounds obey the 18-electron rule. This is well exemplified by the series of complexes $\pi\text{-}C_5H_5Co(CO)_2$, $\pi\text{-}C_5H_5Mn(CO)_3$ and $\pi\text{-}C_5H_5V(CO)_4$. The simplest derivatives of the elements Cr, Fe and Ni are binuclear and held together either by metal-metal bonds or by bridging carbonyl groups and metal-metal bonds. The formation of such binuclear systems by alternate members of the transition series should be compared with the similar formation of dimeric metal carbonyls in the series $Cr(CO)_6$, $Mn_2(CO)_{10}$, $Fe(CO)_5$, $Co_2(CO)_8$ and $Ni(CO)_4$.

It will be seen from Figure 43 that crystalline $[\pi\text{-}C_5H_5Fe(CO)_2]_2$ contains bridging carbonyl groups, whereas the two halves of the molecule $[\pi\text{-}C_5H_5Os(CO)_2]_2$ are held together purely by a 2-electron metal-metal bond; no evidence for any carbonyl bridged isomer has been obtained in the latter case. The ruthenium complex is intermediate in character and an equilibrium between bridged and unbridged forms has been observed in solution from studies of the infrared spectrum in the carbonyl stretching region. In cyclohexane at 30°C the unbridged form accounts for 55% of the whole, whereas for the iron complex only 0·6% of the unbridged isomer is present under similar conditions. Thus the tendency to form structures involving bridging carbonyl groups seems to fall off in the order 1st > 2nd > 3rd transition series. This is also borne out by the structures of $Fe_3(CO)_{12}$ (contains bridging CO groups), and $Os_3(CO)_{12}$ in which the three $Os(CO)_4$ units are held together by metal-metal bonds only. These trends may be due to the increase in the size of the metal atoms with increase in atomic number. The metal-metal distance may become too large for a CO ligand to bridge.

Similar complexes containing bridging nitrosyl, isocyanide, phosphine, and hydride ligands, are known, see Figure 43.

A number of π-cyclopentadienyl metal complexes are known which contain clusters of three or more metal atoms, held together by metal-

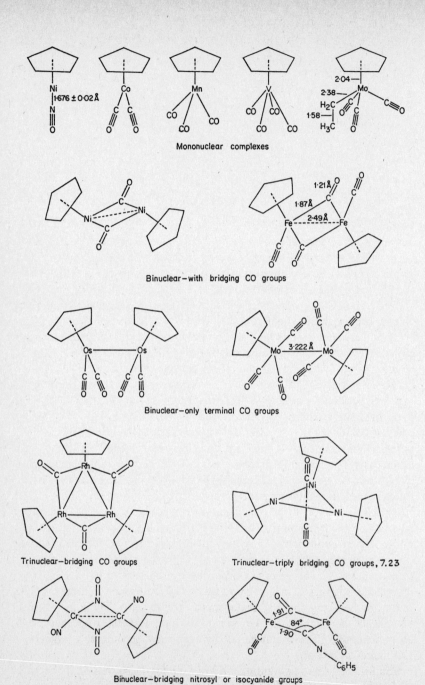

Figure 43. Some π-cyclopentadienyl carbonyl and nitrosyl complexes. Some molecules e.g. [π-C₅H₅Ni(CO)]₂ and [π-C₅H₅Fe(CO)₂]₂ have different structures in solution and in the crystal.

metal bonds and sometimes by bridging ligands as well. Some examples are illustrated (Figure 43). Of particular interest is the nickel complex 7.23, in which the two carbonyl groups each bridge all three metal atoms.

Solution infrared spectroscopic studies of carbonyl complexes are particularly helpful in deciding the types of carbonyl groups which they contain. The carbonyl stretching frequency is related to the order of the CO bond, which clearly falls off as follows: terminal CO groups > doubly-bridging CO > triply bridging CO. Approximate values of the carbonyl stretching frequency to be expected for neutral, uncharged carbonyl complexes of these three types are given in Figure 44:

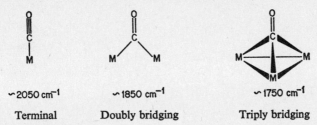

~2050 cm⁻¹	~1850 cm⁻¹	~1750 cm⁻¹
Terminal	Doubly bridging	Triply bridging

Figure 44. The C—O stretching frequency of some metal carbonyl systems.

π-Cyclopentadienyl metal carbonyls are usually rather stable thermally, and in the solid state most resist oxidation by oxygen at room temperature. They can often be reduced by sodium amalgam in tetrahydrofuran to yield π-cyclopentadienyl carbonyl anions. This reduction may proceed either by loss of carbon monoxide, e.g.

$$\pi\text{-}C_5H_5V(CO)_4 \xrightarrow[\text{THF}]{\text{Na/Hg}} (Na^+)_2[\pi\text{-}C_5H_5V(CO)_3]^{2-} + CO$$

or by cleavage of a metal-metal bond, e.g.

$$[\pi\text{-}C_5H_5Fe(CO)_2]_2 \xrightarrow[\text{THF}]{\text{Na/Hg}} 2[\pi\text{-}C_5H_5Fe(CO)_2]^-Na^+$$

$$[\pi\text{-}C_5H_5Mo(CO)_3]_2 \xrightarrow[\text{THF}]{\text{Na/Hg}} 2[\pi\text{-}C_5H_5Mo(CO)_3]^-Na^+$$

Acidification of solutions containing salts of the anions leads to the liberation of the metal hydrides which contain covalent metal-hydrogen bonds, e.g.

$$[\pi\text{-}C_5H_5Mo(CO)_3]^- + H^+ \longrightarrow \pi\text{-}C_5H_5Mo(CO)_3H$$

As mentioned in Chapter 6, treatment of the anions with alkyl halides produces σ-alkyl transition metal derivatives. Other reactions typical of both π-cyclopentadienyl carbonyl anions and carbonyl anions are illustrated in Figure 45.

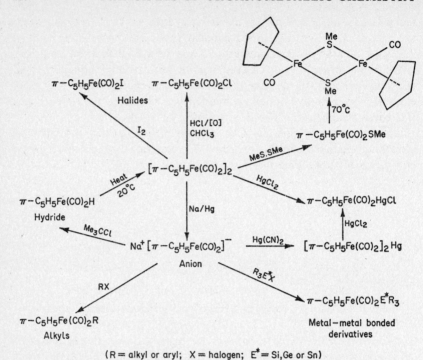

$(R = alkyl \text{ or } aryl; \quad X = halogen; \quad E^* = Si, Ge \text{ or } Sn)$

Figure 45. Some reactions of bis (π-cyclopentadienyliron dicarbonyl).

Tricarbonyl(methylcyclopentadienyl)manganese, π-MeC$_5$H$_4$Mn(CO)$_3$, is the only cyclopentadienyl to have found industrial application on a significant scale. It is used as an antiknock, mixed with tetra-methyl and -ethyl lead and their redistribution products, and in some ways its anti-knock properties are complementary to those of the lead compounds. Its antiknock property is due to the ability of manganese oxides, produced in the combustion, to end radical chains by electron-transfer processes (as PbO does). The methylcyclopentadienyl compound, which is liquid at room temperature, is preferred to π-C$_5$H$_5$Mn(CO)$_3$ (mp 77°) because liquids are more convenient to handle industrially than solids. It is also used as a 'combustion improver' since it helps to reduce smoke.

Six-electron ligands

Arene complexes

The most familiar 6-electron ligands are benzene and substituted benzenes. Ligands of this type are called 'arene'. The general methods of preparation of their complexes have already been described (p 169). Quite a wide range of arene complexes is known, but the diversity of types is somewhat smaller than those formed by the π-cyclopentadienyl ligand. Both bis-π-arene complexes, e.g. $(\pi\text{-}C_6H_6)_2M$ (M = V, Cr) and (π-mesitylene)$_2$M (M = V, Cr, Fe, Co), and cations derived from them, e.g. [(hexamethyl-benzene)$_2$Mn]$^+$ are known, as are mono-π-arene complexes, in which one arene group is present in the molecule together with other ligands, e.g. $\pi\text{-}C_6H_6Cr(CO)_3$.

The neutral bis-π-arene complexes form well defined crystals, moderately soluble in common organic solvents. They may be sublimed in vacuum at about 100°C. Thermally they are reasonably stable, frequently up to 300°C. Some mean bond dissociation energies of the metal-ring bonds are listed in Table XVI together with some equivalent data for bis-π-cyclopentadienyl complexes.

Table XVI. *Some thermodynamic data on bis-π-arene complexes*

Metal-ring bond in the complex	$\bar{D}$ [a] kcal/mole
$(\pi\text{-}C_6H_6)_2V$	70.0 ± 2
$(\pi\text{-}C_6H_6)_2Cr$	40.5 ± 8
$(\pi\text{-Mes})_2Cr$	41.5 ± 2
$(\pi\text{-}C_6H_6)_2Mo$	50.7 ± 2
$(\pi\text{-}C_5H_5)_2Fe$	69.5 ± 10
$(\pi\text{-}C_5H_5)_2Ni$	56.6 ± 10

[a] $\bar{D}$ is the mean bond dissociation energy of MR$_n$;
$\bar{D}$(M—R) = $\Delta H°/n$; see reference to Skinner, p 30.

These data suggest that the metal-ring bond in ferrocene is almost twice as strong as that in bis-π-benzene chromium, the corresponding 18-electron complex of the bis-π-arene series.

Most of the neutral bis-π-arene metal complexes are readily oxidized to bis-arene cations. Indeed, bis-π-benzene chromium acts as an electron donor, for example it forms 1:1 complexes with acceptor molecules such as tetracyanoethylene, which are probably best formulated as salts $[(\pi\text{-}C_6H_6)_2Cr]^+TCNE^-$.

Electron-diffraction studies on bis-π-benzene chromium in the vapour state show it to have the sandwich structure (D$_{6h}$ symmetry) in which all the C—C bond distances are equal (1.423 ± 0.002 Å), and the arene rings are planar and parallel to each other.

The X-ray structures of benzene chromium tricarbonyl and of hexa-methyl benzene chromium tricarbonyl also suggest that the C—C bond distances are equivalent. In the hexamethylbenzene complex the methyl

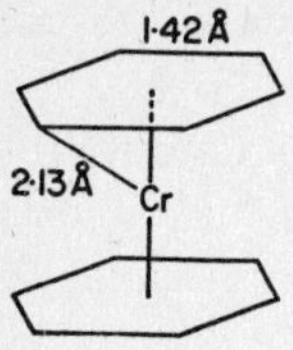

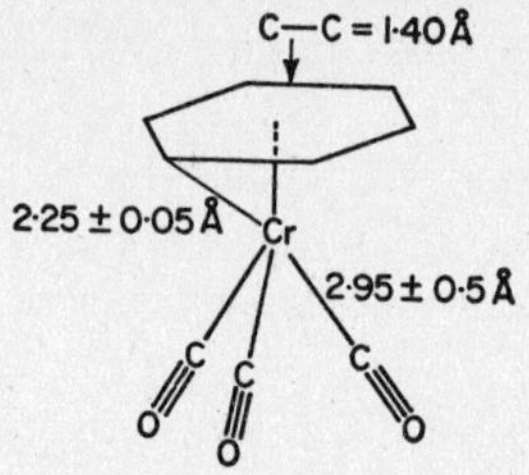

7.24 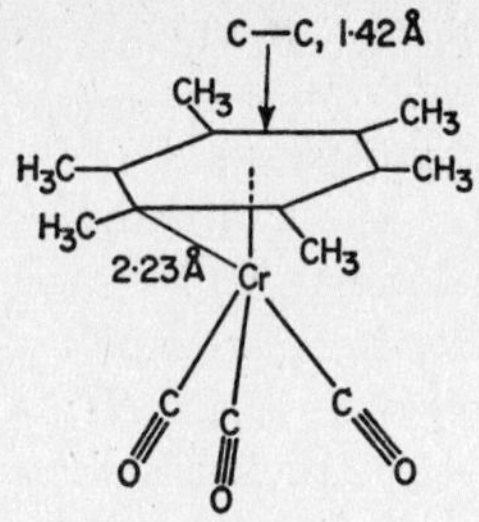　　　　　　　7.25

and ring carbons are essentially coplanar and their plane is parallel to that formed by the three carbons of the CO groups. Both the molecules 7.24, and 7.25 possess the staggered conformation in the crystal.

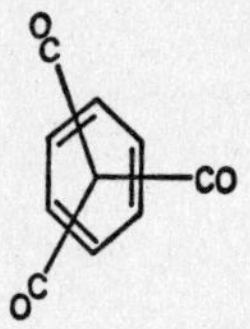

Some of the reactions of arene complexes, notably substitution of the ring, are described in Chapter 8.

Olefin 6-electron ligands

A number of olefins may also act as 6-electron ligands. A cyclo-octa-tetraene molybdenum complex in this category is $C_8H_8Mo(CO)_3$. Similarly with Group VI metal carbonyls, cycloheptatriene forms complexes $C_7H_8M(CO)_3$. The crystal structure of the molybdenum derivative, 7.26, shows that the six sp^2 carbon atoms form a plane, with the molybdenum atom lying below and almost equidistant from all six carbon atoms (2·53 Å). Here, however, the C—C distances clearly alternate as in the free hydrocarbon. The methylene group of the C_7H_8 ligand lies above the C_6-plane which suggests some sp^3 character for the carbon atoms to which it is attached. The structure also shows that the three carbonyl groups are essentially diametrically opposed to the positions of the carbon-carbon double bonds, so that the metal atom possesses an approximately octahedral environment. Cyclo-octatriene chromium tri-carbonyl 7.27, has a similar structure.

This favoured stereochemistry, together with the 18-electron configuration of the molybdenum atom, account for the preference of the cyclohepta-triene to act as a 6-electron ligand in this compound. On the other hand iron often favours a diene ligand rather than a triene; this is probably due to the stability of the $M(CO)_3$ group. Iron therefore forms the cyclo-heptatriene complex $C_7H_8Fe(CO)_3$ where only 4 π-electrons of the triene bond to the iron.

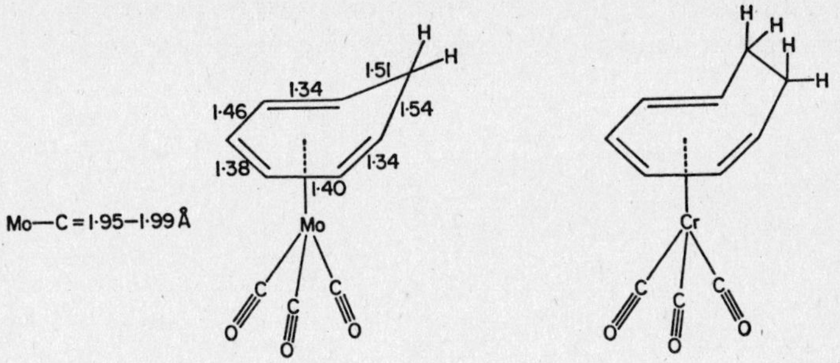

7.26   7.27

Transition metal complexes containing π-bonded heterocyclic ligands

Heterocyclic compounds such as pyrrole or thiophene can also act as 5- or 6-electron ligands to transition metals. Only a few of their complexes are known to date, but there seems no reason why a wide variety of unsaturated heterocyclics should not form stable, π-bonded metal complexes. Some examples, with preparative routes, are given below:

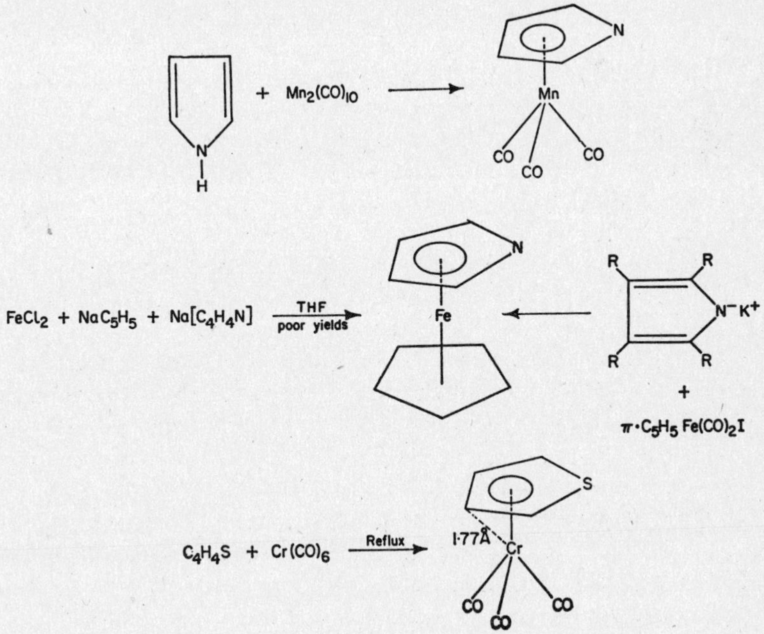

Of particular interest are various complexes formed by organoboron compounds with transition elements. Treatment of the N-lithio-derivative of the 5-membered boron-nitrogen heterocycle, 7.28, with anhydrous ferrous chloride yields a dark-brown diamagnetic product, for which the sandwich ferrocene-like structure 7.29, has been proposed.

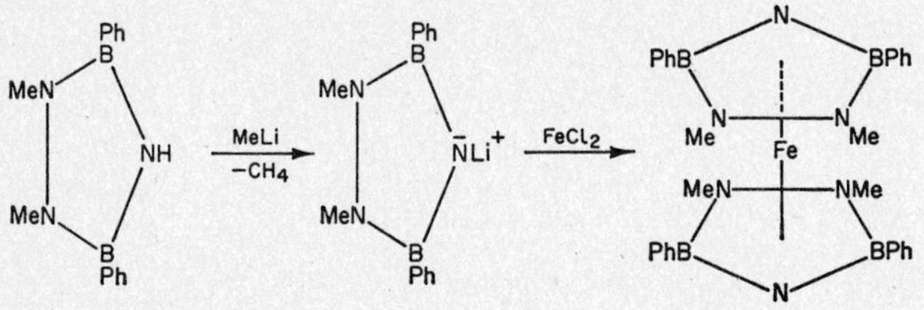

7.28

7.29

A hexamethylborazine chromium tricarbonyl complex is also known. It seems reasonable that it will have an eclipsed configuration with the 'donor' nitrogen in the position *trans* to the 'π-acid' carbonyls. The ring-chromium bonding would be expected to be more localized than in benzeneCr(CO)$_3$, but less so than in a complex L$_3$Cr(CO)$_3$, where L is (for example) pyridine.

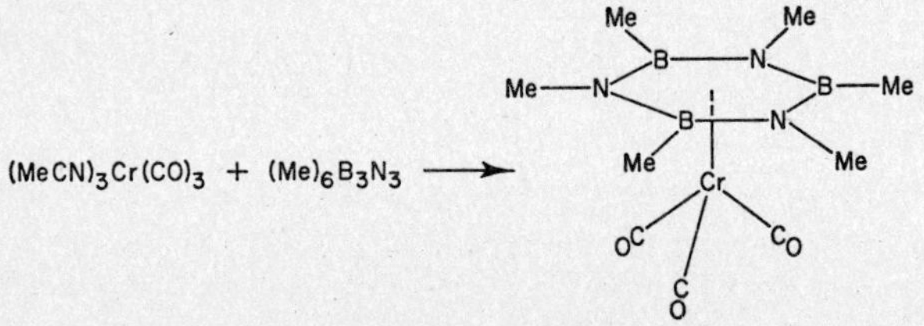

$$(MeCN)_3Cr(CO)_3 + (Me)_6B_3N_3 \longrightarrow$$

Transition metal sandwich and 'half-sandwich' complexes of carborane ligands such as B$_9$C$_2$H$_{11}$ are described in Chapter 3, p 81.

Seven-electron ligands

Cycloheptatriene complexes, in which the olefin is acting as a 6-electron ligand, may undergo hydride abstraction to give π-cycloheptatrienyl metal cations, e.g.

The reverse type of reaction can also occur, when π-cycloheptatrienyl metal cations are treated with anions such as $Y^- = CN^-$, $\bar{O}Me$, $\bar{C}H(COOEt)_2$

In some cases, the expulsion of a hydrogen atom from cycloheptatriene occurs spontaneously, e.g.

$$\pi\text{-}C_5H_5V(CO)_4 + C_7H_8 \longrightarrow \pi\text{-}C_7H_7V\text{-}\pi\text{-}C_5H_5$$

The structure of π-cyclopentadienyl-π-cycloheptatrienyl vanadium, 7.30, has been determined by X-ray diffraction. The C_7H_7 ring is planar and the C—C bonds distances are equal within experimental error.

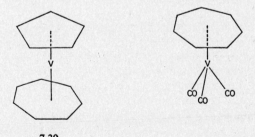

7.30

Figure 46. The structures of π-cyclopentadienyl-π-cycloheptatrienyl vanadium and π-cycloheptatrienyl vanadium tricarbonyl.

P

An example of a ring replacement reaction involving mixed sandwich complexes is,

$$\pi\text{-}C_5H_5\,Cr\,\pi\text{-}C_6H_6 \xrightarrow[\;C_7H_8\;]{AlCl_3} \quad Cr^+ \xrightarrow{\text{Alkali, }Na_2S_2O_4} \quad Cr$$

7.31

It should be noted that the complex, 7.31, is isoelectronic with bis-π-benzene chromium.

It might be expected that azulenes would be able to act as 7-electron ligands. However, in those complexes whose crystal structures have been determined this is not the case. In the iron complex, 7.32, azulene acts as a $5+3$ electron ligand whilst in the molybdenum compound, 7.33, azulene is a 2×5 electron ligand.

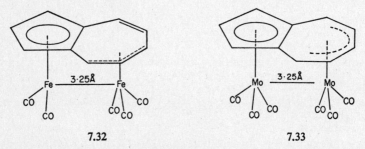

7.32   7.33

Cyclo-octatetraene complexes

As discussed on p 166, cyclo-octatetraene can bond to transition metals in many different ways. The room temperature proton magnetic resonance spectrum of the complex $C_8H_8Fe(CO)_3$ shows only one sharp line. On this basis a planar delocalized C_8 ring structure was first, but incorrectly, proposed for this molecule. Low temperature spectra suggest that in solution the cyclo-octatetraene rings are bound to the metal by 1,3 diene systems as found in the crystal structure, (see Figure 32, p 166). It is thought that the hydrogen atoms of the ring become equivalent in the proton magnetic resonance spectrum due to a rapid circular, essentially 'in plane', movement of the ring about the iron-ring axis.

Cyclo-octatetraene can act formally as an 8-electron ligand, as in the compound $(C_8H_8)_3Ti_2$, 7.34. The external C_8 rings are planar and the annular carbon atoms equidistant from the titanium atom (2·35 Å). The bonding between the metal and the planar rings is probably largely ionic. The dianion $C_8H_8^=$ is known and has a planar structure, as would be expected for an aromatic system with $4n+2$, (10), π-electrons.

7.34

BIBLIOGRAPHY

Structural chemistry of organo-transition metal complexes

M. R. Churchill and R. Mason, *Advances in Organometallic Chemistry*, Volume 5, edited by F. G. A. Stone and R. West (Academic Press, New York and London, 1967), p. 93. A review of some recent developments.

σ-Alkyls and aryls

G. W. Parshall and J. J. Mrowca, *Advances in Organometallic Chemistry* 7, 1968, 157.

Complexes with di- and oligo-olefinic ligands

E. O. Fischer and H. Werner, 'Metal π-Complexes', Volume I (Elsevier, Amsterdam, 1966). A comprehensive account, which also includes a discussion of π-enyl complexes.

π-Allyl complexes

M. L. H. Green and P. L. I. Nagy, *Advances in Organometallic Chemistry* 2, 1964, 325.
R. G. Guy and B. L. Shaw, *Advances in Inorganic Chemistry and Radiochemistry*, Volume 4, edited by H. J. Emeleus and A. G. Sharpe (Academic Press, New York and London, 1962), p. 78. Also olefin and acetylene complexes.

Cyclobutadiene metal complexes

P. M. Maitlis, *Advances in Organometallic Chemistry* 4, 1966, 95.

Diene-iron carbonyl complexes

R. Pettit and G. F. Emerson, *Advances in Organometallic Chemistry* 1, 1964, 1.

Cyclopentadienyl and arene complexes. Reviews of early work

H. P. Fritz and E. O. Fischer, *Advances in Inorganic Chemistry and Radiochemistry* 1, 1959, 55.
G. Wilkinson and F. A. Cotton, *Progress in Inorganic Chemistry* 1, 1959, 1.

Complexes containing 'dynamic' organic ligands.

F. A. Cotton, *Accounts of Chemical Research*, 1, 1968, 257.

P. L. Pauson, in 'Non-Benzenoid Aromatic Compounds', edited by D. Ginsburg (Interscience, London, 1959) and in 'Organometallic Chemistry', edited by H. Zeiss (Reinhold, New York, 1960). Cyclopentadienyls.

H. Zeiss, in 'Organometallic Chemistry', edited by H. Zeiss (Reinhold, New York, 1960). Arene Complexes.

H. Zeiss, P. J. Wheatley and H. J. S. Winkler, 'Benzenoid-Metal Complexes', (The Ronald Press Co., New York, 1966). A review.

Cyclopentadienyl complexes

M. Rosenblum, 'Chemistry of the Iron Group Metallocenes', Part I, (Wiley, New York and London, 1965). A comprehensive text.

π-Complexes formed by Seven-Membered and Eight-Membered Carbocyclic Compounds

M. A. Bennett, *Advances in Organometallic Chemistry* 4, 1966, 353. A detailed review.

CHAPTER EIGHT

The organic chemistry of ferrocene and related compounds

The aromatic character of cyclic C_nH_n ligands in transition metal complexes

According to Hückel's rule, carbocyclic species possessing $(4n+2)$ π-electrons in a delocalized system should exhibit aromatic properties. The C_nH_n species shown in Table XVII should all fall into this category:

Table XVII

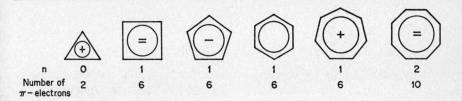

n	0	1	1	1	1	2
Number of π – electrons	2	6	6	6	6	10

One characteristic of an aromatic system derived from carbocyclic species C_nH_n is that the ring should be planar. For all the species in Table XVII (except $C_4H_4^{2-}$), the planarity of the ring has been demonstrated. Moreover, as was pointed out in Chapter 7, this planarity is retained in transition metal complexes derived from these species, where all the annular carbon atoms are involved in bonding to the transition metal. It would therefore be anticipated that, in their complexes, the ligands would continue to show chemical properties characteristic of aromatic compounds, especially the property of undergoing substitution, rather than addition reactions, with electrophiles.

It should perhaps be stressed here that in this book cyclobutadiene complexes have been classified as being derived from the diradical $C_4H_4\cdot\cdot$.

Similarly cyclopentadienyl complexes have been treated as being formed by the radical $C_5H_5\cdot$. In other words, they are considered as acting as 4- or 5-electron ligands respectively when they bond to a transition metal. This classification is really a formality, and the complexes could just as well be thought of as being derived from the 6-electron aromatic anions $C_4H_4^=$ and $C_5H_5^-$ respectively and a metal cation carrying a suitable positive charge ($+2$ or $+1$) to maintain electroneutrality. The total number of electrons in the complex, of course, remains the same.

The cyclopropenyl cation $C_3H_3^+$, should, on the basis of Hückel's rule, also be an aromatic system with 2 π-electrons; the complex [(π-Ph_3C_3)-NiCl(py)$_2$].py, 8.1, has been shown, by X-ray diffraction, to contain the triphenyl-cyclopropenyl ligand. The geometry around the central nickel atom is a distorted tetrahedron.

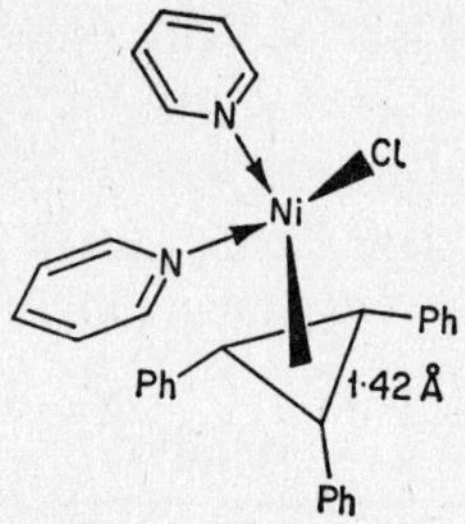

8.1

Aromatic properties of co-ordinated ligands

Woodward and his co-workers observed that ferrocene was not hydrogenated under normal conditions and that it did not undergo the Diels-Alder reaction which is characteristic of conjugated dienes. It could however be acetylated readily under Friedel-Crafts conditions. This revealed the aromatic properties of the π-cyclopentadienyl ring, and subsequently similar aromatic behaviour has been observed for e.g., ruthenocene, osmocene, π-$C_5H_5Mn(CO)_3$ and π-$C_5H_5V(CO)_4$. Indeed, for the most part π-cyclopentadienyl systems may be assumed to be 'aromatic', though particular complexes may not undergo typical substitution reactions since the metal can interfere with the reaction (thus oxidation to a cationic species would inhibit electrophilic attack), or the molecule might not survive the conditions of the reactions.

Electronic effects

The description of the bonding in ferrocene suggests that there is a higher electron density on the π-cyclopentadienyl ring compared with that in benzene. The electron density on the rings in comparable complexes should fall off for the series

$$\pi\text{-}C_4H_4 > \pi\text{-}C_5H_5 > \text{benzene} > \pi\text{-}C_6H_6 > \pi\text{-}C_7H_7$$

and this is borne out in part by the acid strengths of substituted carboxylic acids, which lie in the order ferrocene carboxylic acid < benzoic acid < benzoic acid $Cr(CO)_3$. Again, the particularly high reactivity of ferrocene and other π-cyclopentadienyl compounds towards electrophilic reagents is well demonstrated by the order of reactivity in Friedel-Crafts acylation:

$$\text{phenol} \approx \text{ferrocene} > \text{anisole} > \pi\text{-}C_5H_5Mn(CO)_3 > \text{benzene} > C_6H_6Cr(CO)_3$$

The π-cyclobutadiene ligand is also substituted readily, e.g.

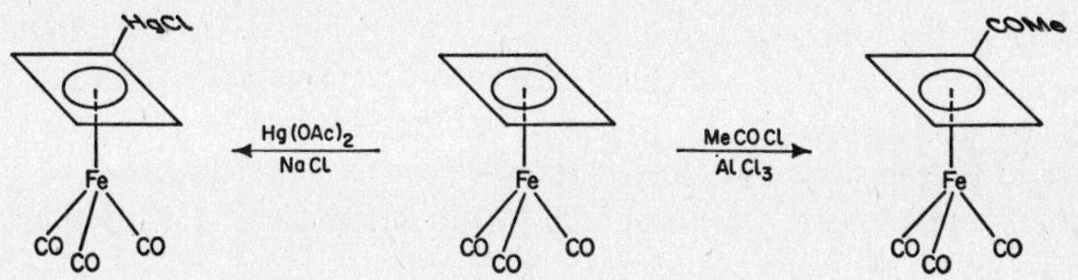

Moreover, acetylation or mercuration of $\pi\text{-}C_4H_4Co\text{-}\pi\text{-}C_5H_5$ results in preferential substitution of the C_4-ring.

In agreement with the above order of reactivities, Friedel-Crafts acylation of $C_6H_6Cr(CO)_3$ does not proceed very easily. The reduced electron density on the arene ring compared with that in the free ligand, however, enhances its susceptibility to nucleophilic substitution:

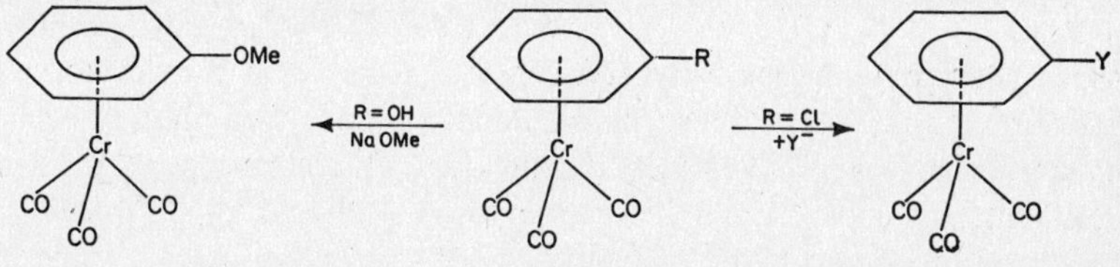

As shown on p 211, nucleophilic attack on $C_7H_7M(CO)_3{}^+$ cations proceeds by addition to yield cycloheptatriene complexes.

In substituted ferrocenes, the substituent affects the reactivity of the substituted ring, the relative reactivity of the two rings, and hence the reactivity of the molecule as a whole. Thus the ease of oxidation of ferrocene is reduced by electron withdrawing, –I, substituents and, in acid media, the relative ease of oxidation is alkylferrocenes > ferrocene > acyl-ferrocene. A striking example is ferrocenyl-carboxaldehyde which resists oxidation by permanganate under conditions which will oxidize benzaldehyde to benzoic acid.

Acylation experiments show that with disubstitution the major product is the 1:1′ diacetyl derivative indicating that, as expected, −I substituents deactivate the substituted ring. In those cases where small yields of products diacetylated in one ring are isolated, the acyl substituents enter the 1,2-positions of the π-cyclopentadienyl ring rather than the 1,3-positions. Thus in the mono-acetylated ring the 3-position is deactivated relative to the 2-position.

Some mechanisms of electrophilic substitution

Apart from the role of bonding with and stabilizing the cyclopentadienyl radicals, the iron atom may also take a direct part in the chemistry of ferrocene substituents, and in the mechanism of substitution. It has been proposed that electrophilic substitutions occur by a general reaction path in which the attacking electrophilic group interacts first with the iron atom to give a π-diene cationic intermediate:

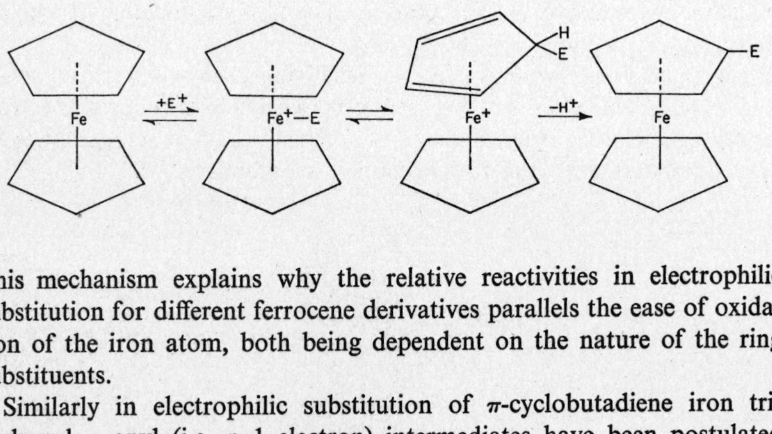

This mechanism explains why the relative reactivities in electrophilic substitution for different ferrocene derivatives parallels the ease of oxidation of the iron atom, both being dependent on the nature of the ring substituents.

Similarly in electrophilic substitution of π-cyclobutadiene iron tricarbonyl, π-enyl (i.e. n−1 electron) intermediates have been postulated cf. p 195, e.g.

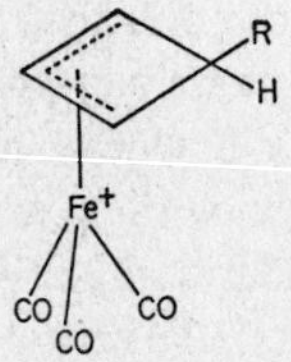

Some particular reactions

Friedel-Crafts acylation, alkylation and related reactions

The Friedel-Crafts acylation of ferrocene, which was mentioned above, has been studied in considerable detail. Substitution can be effected under very mild conditions, for instance with acetic anhydride containing phosphoric acid as catalyst, to give mono-acetylferrocene. With aluminium chloride as catalyst, good yields of either mono- or 1,1'-diacetylferrocene can be obtained according to the proportions of reagents taken.

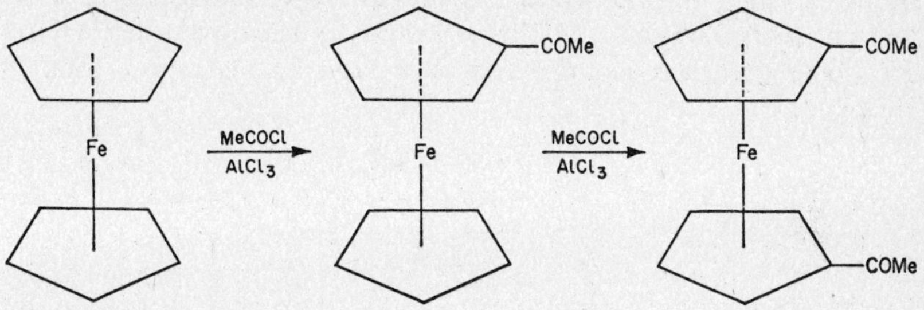

Ferrocene also undergoes the Vilsmeier reaction to yield ferrocene-carboxaldehyde, which is a useful starting material for the synthesis of other ferrocene derivatives, for example by condensation with compounds containing active CH_2 groups such as malonic acid:

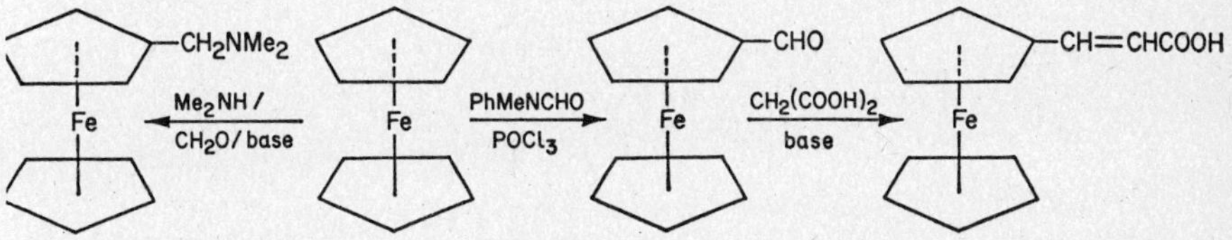

8.2

With formaldehyde and secondary amines, (Mannich condensation) ferrocene yields mono-dialkylaminomethyl derivatives, 8.2. The ability of ferrocene to react under Vilsmeier or Mannich conditions is further evidence of its high susceptibility to electrophilic substitution, as these reactions are observed in the benzene series only for reactive compounds such as phenol.

Direct Friedel-Crafts alkylation of ferrocene gives only low yields of mono- and poly-alkylated ferrocenes. Better yields are obtained using aluminium chloride and olefins under pressure:

$$(\pi\text{-}C_5H_5)_2Fe + CH_2{=}CH_2 \xrightarrow[\text{30–40 atm,. 100–130°}]{\text{AlCl}_3} \pi\text{-}C_5H_5Fe\text{-}\pi\text{-}C_5H_4CH_2CH_3$$
$$\underset{\text{+ polyalkylated ferrocenes}}{\sim 20\%}$$

Dialkylation occurs in the same ring preferentially, as would be expected on account of the electron donating ($+I$) effect of alkyl substituents. This explains the large proportion of polyalkylated products obtained in Friedel-Crafts alkylations. Monoalkylated ferrocenes are best prepared by indirect routes, such as by Clemmensen reduction of acylferrocenes:

$$\pi\text{-}C_5H_5Fe\text{-}\pi\text{-}C_5H_4COPh \xrightarrow{\text{Zn/HCl}} \pi\text{-}C_5H_5Fe\text{-}\pi\text{-}C_5H_4CH_2Ph$$

Nitration and Halogenation

Direct nitration by nitric acid mixtures or direct halogenation using halogens leads in the case of ferrocene either to oxidation or decomposition. Indirect routes are therefore necessary; examples are given in Figure 47.

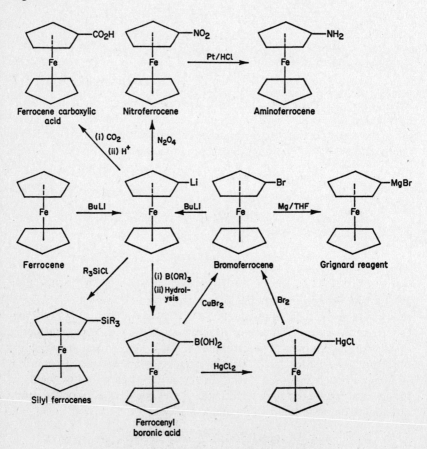

Figure 47. Some reactions of ferrocenyl lithium.

Metalation

The π-cyclopentadienyl rings of ferrocene are readily metalated; for example by treatment with butyl-lithium. This metalation is similar to the hydrogen/metal exchange reactions typical of, for example, certain benzenoid hydrocarbons. (p 48). Mixtures of mono- and di-metalated products are frequently obtained. A suitable method of preparing 1,1'-dilithioferrocene is given on p 49.

Halogenated ferrocenes give rise to pure metalated products by the halogen/metal exchange reaction with butyl lithium (see Figure 47). The metalated derivatives are extremely sensitive to oxidation and hydrolysis and further reactions are normally carried out at once. They are versatile reagents for the synthesis of substituted ferrocenes. Some typical applications are listed in Figure 47. These may be compared with similar reactions of organolithium compounds described in Chapters 2 and 3.

The hydrides $(\pi\text{-}C_5H_5)_2\text{ReH}$ and $(\pi\text{-}C_5H_5)_2\text{WH}_2$ react with excess butyl-lithium to give 1,1'-lithiated derivatives. The di-lithiated rhenium complex reacts in a most unusual way with methyl iodide to give the methyl complex, 8.3, whose structure has been determined by X-ray analysis. The mechanism of the reaction is unknown, but possibly involves *inter alia* the transfer of a methyl group from a ring to the metal.

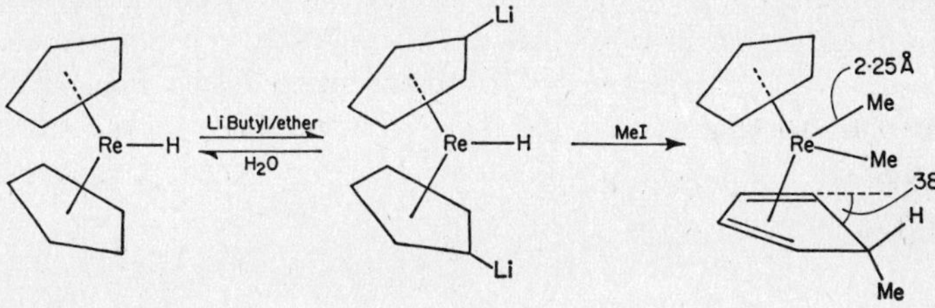

8.3

The interaction of the iron atom with ring substituents; α-carbonium ion stabilization

A dominant feature of the chemistry of ferrocenes is the ease of stabilization of carbonium ions in which a positive charge is localized at the α-position. For example a vinyl substituent in ferrocene is readily protonated, even by acetic acid:

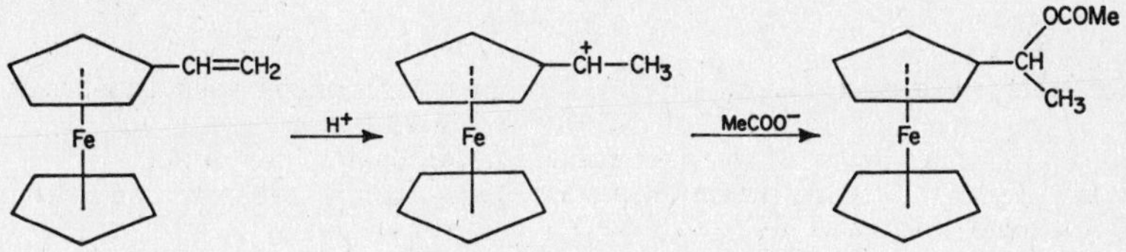

Evidence that the iron atom interacts directly with the α-position is obtained from the observation that the *exo*-acetate, 8.4, is solvolyzed 2500 times faster than the *endo*-acetate, 8.5. In both cases the product is exclusively the *exo*-alcohol.

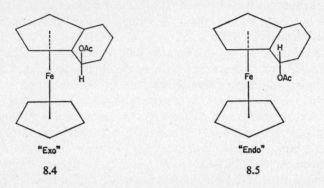

To explain the stabilization of these carbonium ion intermediates it has been suggested that there may be delocalization of the ring electrons to the vacant *p*-orbital of the carbonium ion, or that there may be direct interaction of this *p*-orbital with the non-bonding orbitals on the iron atom (see Figure 48). A more extreme interaction between the ring and the α-carbon than shown in (*b*) could lead to a structure better represented by the *exo*-cyclic structure (*c*). All these mechanisms may occur to some extent and can account for the stereo-specificity of the solvolysis of the isomers 8.4 and, 8.5.

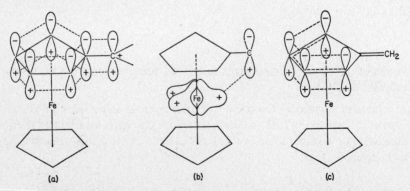

Figure 48. Possible mechanisms for the stabilization of carbonium ions in the α-position. on the π-C_5H_5 ring.
    (*a*) Delocalization of ring π-electrons to the carbonium ion *p* orbital.
    (*b*) Direct interaction between metal lone pairs and the carbonium ion *p* orbital.
    (*c*) *exo*-Cyclic structure.

The analogy between these iron-carbonium ion interactions in ferrocene and similar interactions believed to occur between metals and the β-position of σ-bonded organic systems discussed in Chapter 7 will be apparent.

Similar β-carbonium ions have been proposed to explain kinetic measurements on the solvolysis of the arene chromium tricarbonyl complexes, 8.6. Thus while the $Cr(CO)_3$ group may withdraw electrons from the arene ring the metal atom can also provide electrons for the stabilization of carbonium ions which are ring substituents.

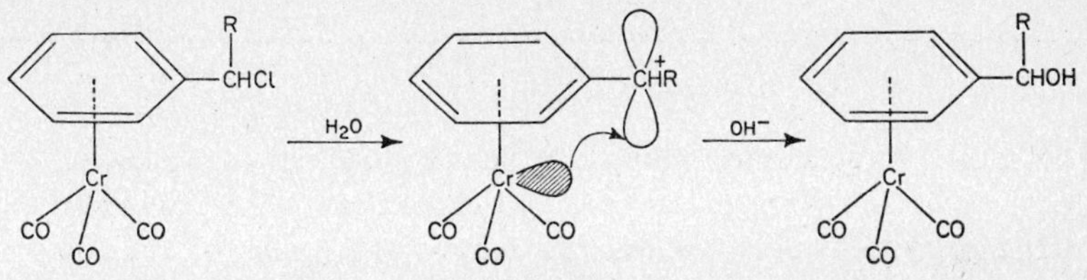

8.6

It is beyond the scope of this book to describe further the organic chemistry of ferrocene. There is much research interest at present in this field. The stability of the ferrocenyl group and the diversity of its reactions permit the preparation of a vast range of derivatives, although no major applications for them have yet been found.

BIBLIOGRAPHY

M. Rosenblum, 'The chemistry of the iron group metallocenes', John Wiley, New York and London, 1965. A good, comprehensive text.
J. P. Collman in *Transition Metal Chemistry*, Volume 2, Ed. R. L. Carlin, Edward Arnold, London, 1966. Reactions of co-ordinated ligands.
K. Plesske, *Angew. Chem.* (International edition), 1962, **1**, 312, 394. Useful reviews, but now slightly out of date.
G. E. Coates, M. L. H. Green and K. Wade, 'Organometallic Compounds', Volume II, Methuen, London, 1968, Chapter 4. A brief summary of important aspects.

CHAPTER NINE

Organometallic complexes formed from acetylenes

The sticky, brown soups formed in the reactions between acetylenes and metal carbonyls yield an amazing variety of complexes including many of the organometallic *exotica*. The nature and yields of the complexes depend markedly on the reaction conditions, in particular the solvent used and the temperature. Similarly the nature of the products varies considerably with changes from one acetylene to another or with a change of stoicheiometry of the metal carbonyl complex. Thus $Fe_3(CO)_{12}$ yields a wider variety of complexes than $Fe(CO)_5$. Any particular reaction may produce some dozen or so isolable organometallic complexes, to say nothing about the purely organic products which are also formed. Progress in this field has been due largely to the use of chromatography to separate products and of X-ray diffraction to find the structures of the often unusual bi- or poly-nuclear organometallic products.

Many of the complexes formed in these reactions may be classified as complexes containing n-electron ligands. They are described separately here, however, because a discussion of their chemistry can, to some extent, provide insight into the manner of their formation. In Table XVIII are shown the skeletal frameworks of the classes of complexes most commonly found in the dimerization and trimerization reactions of acetylenes. The Table also exemplifies the classification used in this chapter.

Mono-acetylene complexes

Mononuclear complexes; acetylenes as 2-electron ligands

In the earlier discussion of the ethylene-metal bond (Chapter 5) the ethylene was compared, in extreme examples, to the carbon monoxide ligand or to two alkyl groups. In the former situation the ethylene behaves as an essentially monodentate ligand whilst in the latter it behaves as a bidentate ligand. Acetylenes are similar in their behaviour.

Table XVIII.

	Number of acetylenes	Number of CO incorporated in ligand		
		0	1	2
MONONUCLEAR	2	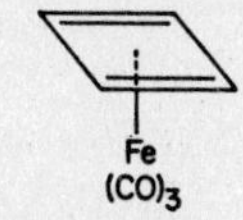 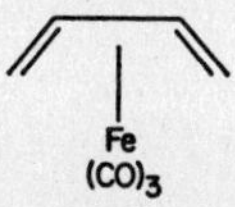 Fe(CO)₃ Cyclobutadiene / Fe(CO)₃ Butadiene	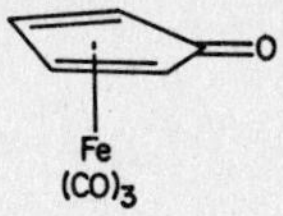 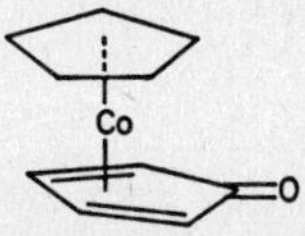 Fe(CO)₃ / Co Cyclopentadienone	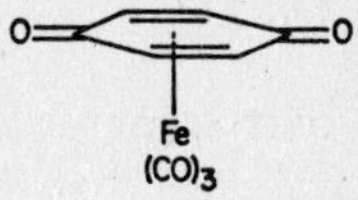 Fe(CO)₃ Quinone
MONONUCLEAR	3	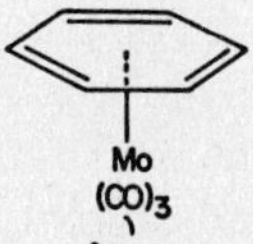 Mo(CO)₃ Arene	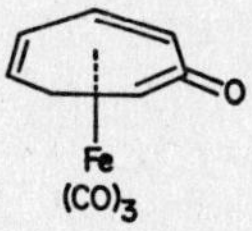 Fe(CO)₃ Tropone	
BINUCLEAR	2	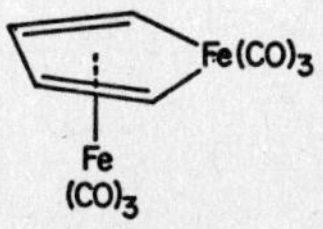 Fe(CO)₃ Fe(CO)₃ Cycloferradiene		
BINUCLEAR	3	(CO)₃Co—Co(CO)₃ / Fe(CO)₃ Fe(CO)₄		

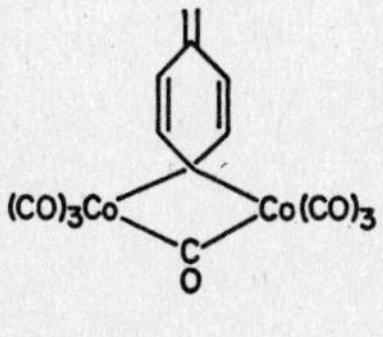

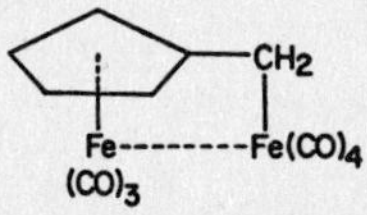

Treatment of π-$C_5H_5Mn(CO)_3$ with acetylenes under irradiation with ultraviolet light gives monoacetylene complexes, e.g.

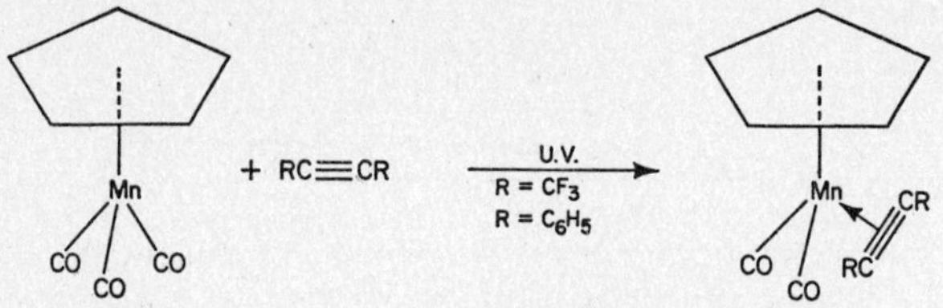

Ethylenic complexes analogous to these are known and it seems reasonable to suppose that the acetylene-metal bond should resemble the ethylene-metal bond in this case. This means that the acetylene, which has two orthogonal $p\pi$-bonds at right angles to each other, would use one filled $p\pi$-bond to donate to the manganese and back-donation from the metal would take place mainly into the corresponding anti-bonding $p\pi^*$-molecular orbital of the acetylene.

Aryl acetylenes form a series of stable complexes of the stoicheiometry (acetylene)$Pt(PPh_3)_2$. In the infrared spectra of these complexes the $C\equiv C$ stretching frequency is lowered to $1750 cm^{-1}$ ($\nu C\equiv C$ in $HC\equiv CH$ is $2100 cm^{-1}$; $\nu C=C$ in $H_2C=CH_2$, $1623 cm^{-1}$). The X-ray analysis of the complex 9.1, (R = Ph) shows a planar structure in which the C—C bond distance of the acetylene group is 1.32 Å, i.e. characteristic of a $C=C$ double bond ($C=C$ in ethylene = 1.33 Å).

$$(Ph_3P)_2PtCl_2 + N_2H_4 . H_2O + EtOH + olefin \longrightarrow olefin\ Pt(PPh_3)_2$$

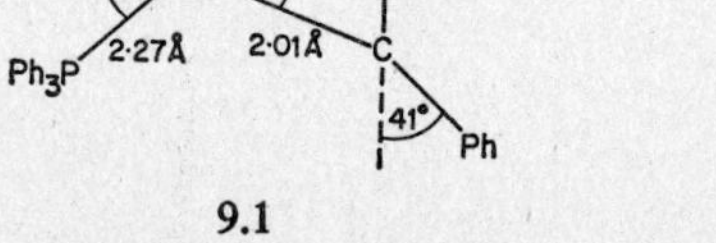

acetylene $RC\equiv CR$

9.1

The phenyl groups are bent 40° out of the C—C axis, which also lends support to the postulate of near sp^2 character for the ligand carbon atoms.

In agreement with this, displacement studies show the order of stability of these complexes to be C_2H_2 < alkylacetylenes < arylacetylenes < nitro-arylacetylenes. Thus the most stable complexes are formed by acetylenes with electron-withdrawing substituents. The presence of such substituents in the acetylene ligand enhances its π-acceptor power, and leads to a stronger metal-ligand π-bond. By analogy with the olefin-platinum complexes (p 160) this structure suggests that π-bonding of the acetylene with the metal is more important than σ-bonding in these compounds.

(ii) Binuclear complexes: acetylenes as four-electron ligands

Treatment of the complexes $Co_2(CO)_8$ and $[\pi\text{-}C_5H_5Ni(CO)]_2$ with acetylenes displaces the bridging carbonyl groups forming derivatives containing bridging acetylene ligands.

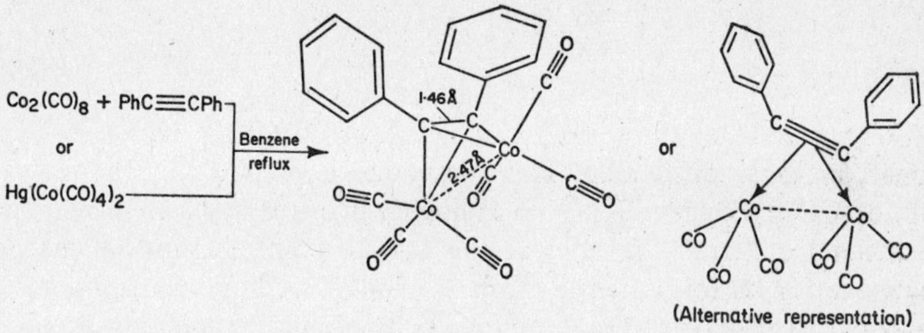

9.2   9.3

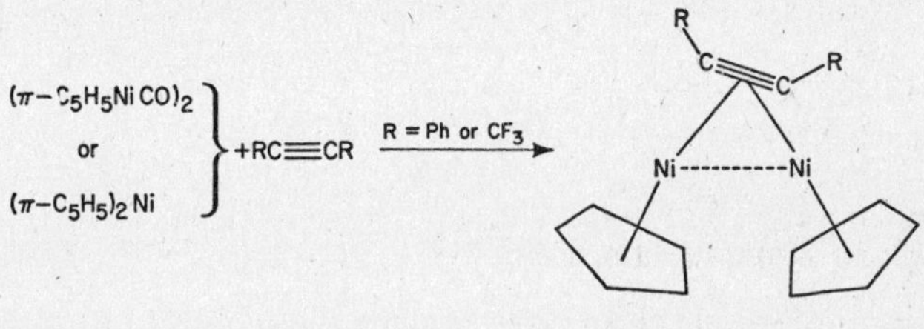

(Alternative representation)

In the cobalt complex the acetylenic carbon atoms are placed between the two cobalt atoms so that the two π-bonds of the acetylene (which are normal to each other) each interact with one cobalt atom. Thus the acetylene may be regarded either as forming two 'π-ethylenic' bonds, one to each cobalt atom, or it can be argued that the acetylenic carbons have become essentially sp^3 hybridized and are forming two σ-bonds to both metals; representations of both points of view are shown in 9.3 and 9.2 respectively.

Trinuclear complexes

Treatment of the bridging acetylene complexes $RC{\equiv}CH(Co(CO)_3)_2$ with hydrochloric acid in methanol gives complexes of stoicheiometry $RC_2H_2 Co_3(CO)_9$. Bromination of these derivatives affords inter alia $RCH_2{-}CBr_3$.

Q

This observation, together with spectral studies, is consistent with the structure, 9.4.

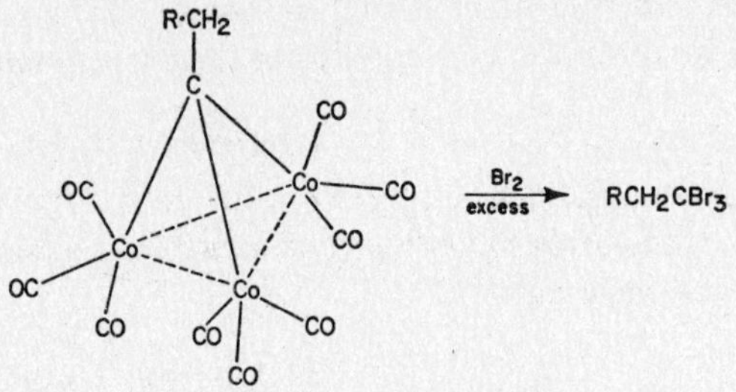

9.4

The structure of the related complex $MeCCo_3(CO)_9$, has been determined by X-ray diffraction. This compound is prepared by the reaction of $Co_2(CO)_8$ with $MeCX_3$ (X = Halogen) in a weakly basic solvent such as ethanol or tetrahydrofuran at the boiling point. A possible reaction sequence for its formation involves the initial disproportionation of the carbonyl to give $[Co(CO)_4]^-$ anions. This reaction commonly occurs in the presence of nitrogen or oxygen bases:

$$3\ Co_2(CO)_8 + 12\ base \longrightarrow 2\ [Co(base)]_6^{2+} + 4[Co(CO)_4]^-_2 + 8CO$$

$$3\ [Co(CO)_4^-] + MeCX_3 \longrightarrow MeC[Co(CO)_4]_3 + 3X^-$$

$$MeC[Co(CO)_4]_3 \longrightarrow MeCCo_3(CO)_9 + 3CO$$

Bis-acetylene mononuclear complexes

The treatment of metal carbonyls (especially those of iron) with acetylenes affords, together with other products, cyclopentadienone and quinone complexes, in which one or two carbonyl groups respectively have been incorporated with two acetylene molecules into a cyclic organic π-bonding system. These complexes can also be obtained by direct reaction between iron carbonyls and cyclopentadienone or quinone ligands. Other products in these reactions include complexes of substituted cyclobutadienes, and π-cyclopentadienyl derivatives. Some examples are shown in Figure 49.

Among the more unexpected complexes formed in the reaction of iron carbonyls with substituted acetylenes are those containing the ferracyclopentadiene system. The crystal structure of one such derivative, 9.5, which was prepared from aqueous $NaHFe(CO)_4$ and 2-butyne, shows one $Fe(CO)_3$ group is incorporated into a 5-membered diene ring and that this ring also acts as a 4-electron ligand to a second $Fe(CO)_3$ system. The

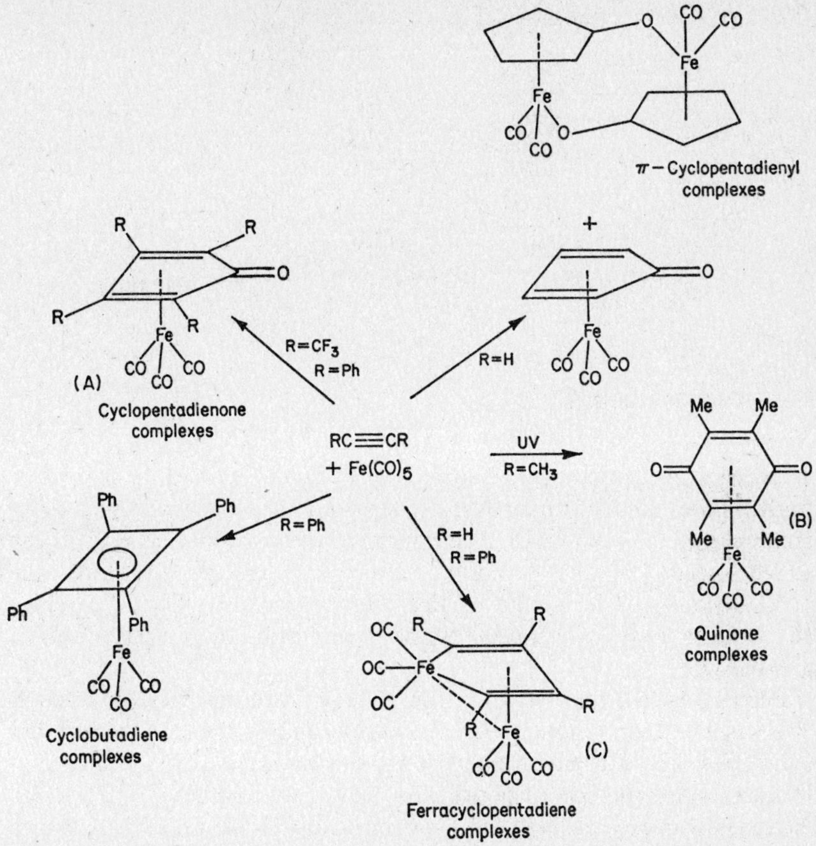

Figure 49. Complexes from iron pentacarbonyl and two acetylene molecules.

A) Also from [diagram] $=O + Fe_2(CO)_9$ (B) Also from $O=$ [diagram] $=O + Fe(CO)_5$

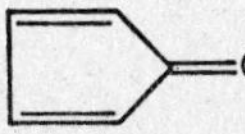

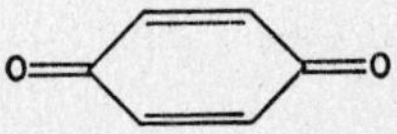

C) Similar complexes from [diagram with S] $+ Fe(CO)_5$ or $NaHFe(CO)_4 + EtC \equiv CH$

Fe—Fe distance in this compound (2·49 Å) is similar to that found in $Fe_2(CO)_9$ (2·49 Å).

This suggests that there is a covalent Fe—Fe interaction between the two iron atoms. Some interaction between the carbon monoxide ligand A and the iron atom of the ferracyclopentadiene ring above it is also indicated.

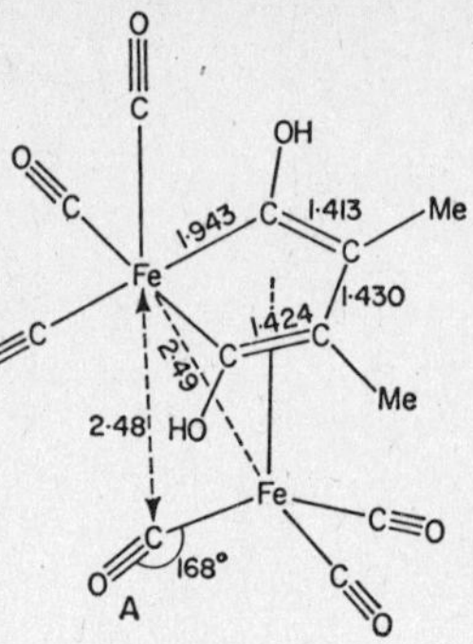

9.5

Tris-acetylene complexes

Mononuclear

Acetylenes may frequently be trimerized by treatment with triphenyl- or trialkyl-chromium compounds to give substituted benzenes and bis-π-arene chromium complexes. The course of the reaction depends on the stoicheiometry of the reaction mixture. Figure 50 illustrates the possible mode of formation of the benzene derivatives. The complex $Me_2Co(PPh_3)\pi$-C_5H_5 can catalyze the trimerization of acetylene itself to give benzene.

Phenylacetylene and $Fe_3(CO)_{12}$ in inert hydrocarbon solvents at 80°C give, as the main products isomeric triphenyltropone iron tricarbonyl complexes. The triphenyltropone ligand can be derived from three acetylene molecules and one carbonyl group (cf. the formation of cyclopentadienone complexes p 229). The X-ray structure of one isomer 9.6, shows that the tropone ring is attached to the iron atom by a diene system, the third double bond being bent out of the plane of the diene carbon atoms, away from the iron (cf. structure of $C_8H_8Fe(CO)_3$ p 166).

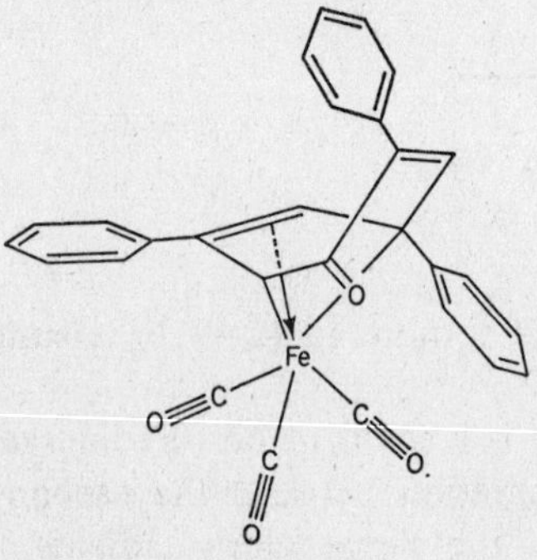

9.6

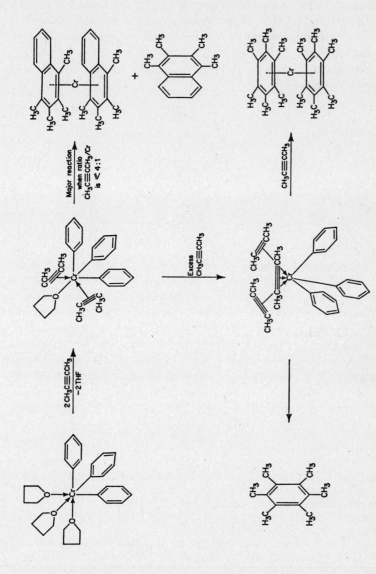

Figure 50. Proposed mechanism of formation of arene-chromium derivatives from acetylenes and chromium-aryl complexes.

Binuclear

Treatment of dicobalt octarbonyl with a 2:1 mixture of *t*-butylacetylene and acetylene forms a remarkable cobalt complex of stoicheimometry $Co_2(CO)_4(C_2HBu^t)_2(C_2H_2)$. The structure of this compound, 9.7, provides a fascinating insight into a mechanism of polymerization of acetylenes. The bonding of the hydrocarbon residue in complex, 9.7, may be described in terms of a 'fly-over', bis-enyl system. The distance $Co-C_3$ (π-enyl) is 2·04 Å.

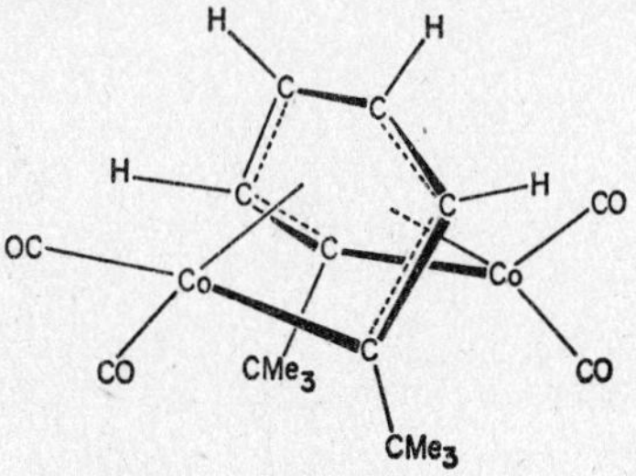

9.7

The complex gives *ortho*-di-*t*-butyl benzenes on decomposition by bromine. The formation of the *ortho* substituted *t*-butyl benzenes which *cannot* be prepared by direct substitution of benzene, can be understood in the light of the structure of the cobalt complex, 9.7, since the carbons attached to the *t*-butyl groups occupy *adjacent* positions in the complex.

The above examples should illustrate the great variety of metal complexes and organic compounds which result from the reaction of acetylenes with compounds of the transition metals. It can be anticipated that reactions of this kind will become increasingly important as synthetic routes to organic compounds which are not otherwise so readily obtained.

BIBLIOGRAPHY

G. E. Coates, M. L. H. Green and K. Wade, 'Organometallic Compounds', Volume II (Methuen, London, 1968), Chapter 8. A similar but more detailed account containing references to the original literature.

G. N. Schrauzer, *Advances in Organometallic Chemistry* 2, 1964, 1. Contains a comprehensive account of nickel acetylene chemistry.

E. O. Fischer and H. Werner, 'Metal-π-Complexes', Volume I (Elsevier, Amsterdam, 1966). Contains a detailed survey of olefin complexes prepared from acetylenes.

M. A. Bennett, *Chem. Rev.* 62, 1962, 611; R. G. Guy and B. L. Shaw, *Advances in Inorganic Chemistry and Radiochemistry* 4, 1962, 78. Slightly out-of-date reviews of olefin and acetylene complexes of transition metals.

F. L. Bowden and A. B. P. Lever, *Organometallic Chemistry Reviews* 3, 1968, 227. Review of complexes derived from acetylenes.

CHAPTER TEN

The role of organotransition metal complexes in some catalytic reactions

In this chapter a number of organic reactions are discussed which involve as catalysts stable, isolable organometallic complexes of the transition metals, or which are thought to involve unstable organometallic intermediates. The polymerization of ethylene by Ziegler catalysts is discussed on p 95. Many of the catalytic reactions discussed below involve the transfer of hydrogen and it is thought that the active catalytic intermediates may contain metal-hydrogen bonds.

Several hundred stable transition metal hydrides are known, in which one or more hydrogen atoms are attached to the metal by essentially covalent bonds.* Frequently, kinetically stable hydrides can be isolated which are analogous to σ-alkyls, that is, they often obey the 18-electron rule, or, in the case of square planar d^8 Rh(I), Ir(I) and Pt(II) are 16-electron compounds. Some typical hydride complexes are shown in Figure 51.

An important property of transition metal hydride complexes in relation to their role as possible catalysts is their ability to add to olefins, often reversibly. This might be compared with the formally similar reaction of boron, aluminium and gallium hydrides which was described in Chapters 2 and 3.

$$M-H + \ \underset{/}{\overset{\backslash}{C}}=\underset{\backslash}{\overset{/}{C}} \ \rightleftharpoons \ M-\overset{|}{\underset{|}{C}}-\overset{|}{\underset{|}{C}}-H$$

Olefin isomerization

Olefins often undergo isomerization by solutions of transition metal complexes. In such isomerizations it is thought that the olefin reacts reversibly with the transition metal to give an organometallic intermediate,

*References to recent reviews are given at the end of this chapter.

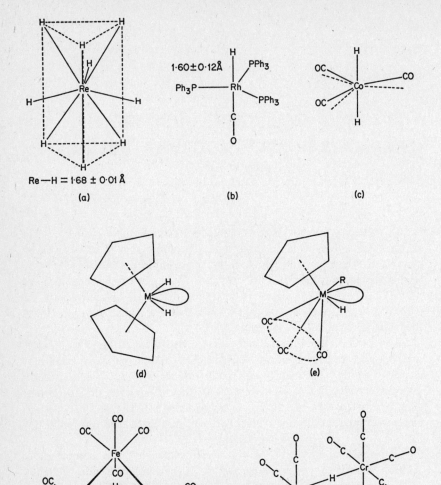

Figure 51. Some transition metal hydride complexes.
- (a) the enneahydrido-rhenate anion $[ReH_9]^{2-}$.
- (b) and (c) trigonal bipyramidal phosphine and carbonyl hydrides.
- (d) bis-π-cyclopentadienyl dihydrides showing lone pair, M=Mo or W.
- (e) proposed structure of π-cyclopentadienyl tricarbonyl alkyl hydride cations, e.g. $[\pi\text{-}C_5H_5Mo(CO)_3MeH]^+$. The circular dotted line shows the plane through the three CO groups and hence analogy of this structure with (d).
- (f) and (g) structures of the anions $[Fe_3(CO)_{11}H]^-$ and $[Cr_2(CO)_{10}H]^-$, showing bridging hydrogen atoms.

in which a hydrogen attached to an sp^3 carbon in the β-position becomes labilized by interaction with the metal atom (see p 188). Two of the most likely mechanisms of isomerization are shown in Figure 52.

In order to act efficiently as a catalyst, the transition metal complex should either contain ligands which can readily be displaced by the incoming olefin, or be co-ordinatively 'unsaturated' (i.e. possess less than a 'stable' e.g. 18-electron configuration) so that the olefin can co-ordinate with the vacant site on the metal. Thus metal carbonyls, e.g. $Fe(CO)_5$ or aqueous solutions of metal salts, e.g. $RhCl_3$, are frequently effective in olefin isomerizations, as the CO and H_2O ligands are often easily displaced to allow the olefin to form a complex. The metal-olefin interaction must also be *kinetically labile*, as unless a complex is formed in which the co-ordinated olefin is readily displaced by other olefin molecules the reaction will be slow or cease altogether owing to 'poisoning' of the catalyst. (see p 237). The rapid isomerization of allyl alcohol to propional-dehyde by $HCo(CO)_4$ is a specific example which illustrates these points. The following mechanism has been proposed:

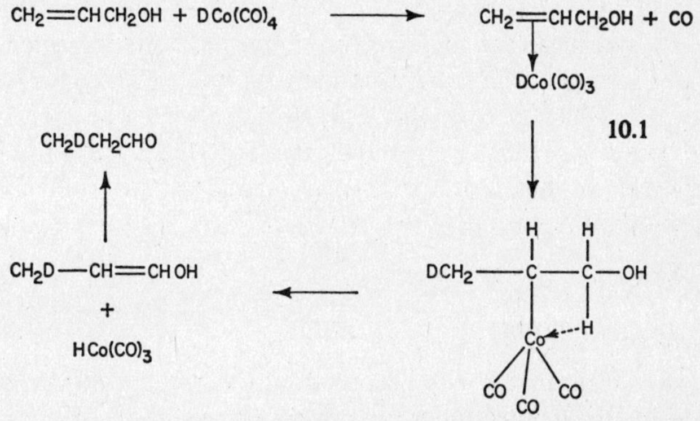

$$CH_2{=}CHCH_2OH + DCo(CO)_4 \longrightarrow CH_2{=}CHCH_2OH + CO$$

$$DCo(CO)_3$$

10.1

$$CH_2DCH_2CHO$$

$$CH_2D{-}CH{=}CHOH$$
$$+$$
$$HCo(CO)_3$$

$$DCH_2{-}\overset{H}{\underset{}{C}}{-}\overset{H}{\underset{}{C}}{-}OH$$
$$\underset{CO\ \ CO}{\overset{}{Co}}{-}H$$
$$CO$$

The first step involves the displacement of one carbon monoxide ligand from $HCo(CO)_4$ by allyl alcohol to form a π-olefin complex, 10.1. Hydrogen is then transferred by a 1,2 shift. Evidence for this mechanism is obtained from studies using the deuteride $DCo(CO)_4$, when $DH_2CCH_2 CHO$ is formed exclusive of $H_3CCHDCHO$ or H_3CCH_2CDO. The reactive intermediate, $HCo(CO)_3$, which contains a free co-ordination position can then add another molecule of allyl alcohol to reform complex 10.1.

Very often isomerization leads to the isomer or mixture of isomers which correspond to the thermodynamically most stable state. Thus

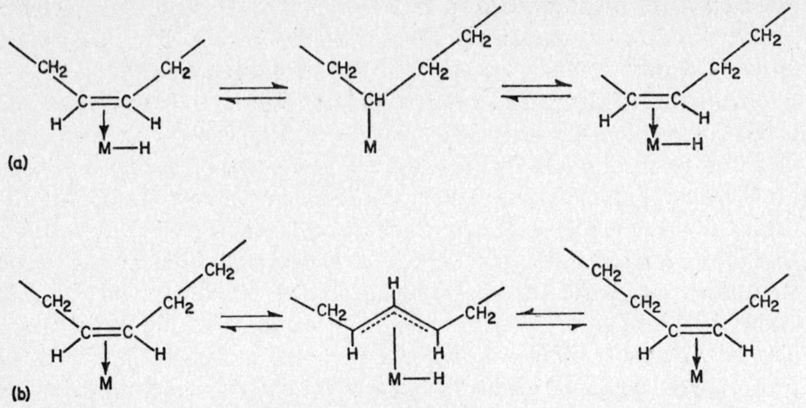

Figure 52. Two mechanisms of isomerization of olefins by hydrogen transfer,
(a) involving 1- to 2-electron ligand equilibria.
(b) involving 2- to 3-electron ligand equilibria.

terminal olefins, e.g. 1-dodecene are converted to internal isomers by $Fe(CO)_5$. (This is the opposite to the direction of isomerization of olefins by boron hydrides, see p 73. Non-conjugated dienes often yield the conjugated isomers, which are usually the more thermodynamically stable. As shown in Chapter 6, p 180, $Fe(CO)_5$ isomerises *cis*-substituted conjugated dienes to *trans*-substituted products. This suggests that both steric and electronic requirements of the metal catalyst may be important in determining the nature of the products.

Homogeneous hydrogenation of olefins

Many transition metal complexes catalyse the hydrogenation of unsaturated hydrocarbons under homogeneous conditions. For example, the complex $(Ph_3P)_3RhCl$, which is prepared as deep violet crystals by heating $RhCl_3.3H_2O$ in ethanol with excess triphenyl phosphine, catalyses the rapid homogeneous hydrogenation of olefins (except ethylene) in benzene solution. Molecular weight determinations show that the complex dissociates in benzene. As the solutions are non-conducting ionization is excluded; the low molecular weight values are attributed to loss of one or more phosphine ligands:

$$Solvent + (Ph_3P)_3RhCl \rightleftharpoons (Ph_3P)_2RhCl(Solvent)_x + Ph_3P$$

The rhodium species in solution is either $(Ph_3P)_2RhCl$, or, more likely, a solvated species $(Ph_3P)_2RhCl(solvent)_x$. The complex reacts with molecular hydrogen in chloroform by displacement of the weakly bound solvent ligands, and a five-co-ordinate dihydride $(Ph_3P)_2RhClH_2$ can be isolated. The benzene solutions also absorb hydrogen rapidly and reversibly. With ethylene the red-brown solutions of $(Ph_3P)_2RhCl(solvent)$ rapidly turn yellow and a crystalline complex $(Ph_3P)_2RhClC_2H_4$ can be isolated. Equilibrium studies in benzene show that under one atmosphere of ethylene more than 90% of the rhodium is in the form of the ethylene complex and proton magnetic resonance data show that the residence time of the ethylene on the rhodium is less than 10^{-2} seconds. In contrast to ethylene, propene forms no isolable complex and equilibrium measurements suggest that only very weak complex formation occurs.

Kinetic measurements under the conditions of the hydrogenation reactions indicate the equilibria (1) and (2):

$$
\begin{array}{c}
\qquad\qquad\qquad \overset{H_2}{\underset{(1)}{\rightleftharpoons}} \\[-4pt]
(PPh_3)_2RhCl(solvent)_x \quad\qquad\qquad (PPh_3)_2RhH_2Cl \\
\qquad\qquad\qquad\qquad\qquad\qquad\qquad\qquad 10.2
\end{array}
$$

$$
\text{olefin} \Big\Updownarrow (2) \qquad\qquad\qquad\qquad \Big\downarrow \text{olefin}
$$

$$
(PPh_3)_2Rh(olefin)Cl \xrightarrow{H_2} (PPh_3)_2RhCl + paraffin
$$

$$
10.3
$$

It is suggested that the catalytic reduction of the olefin proceeds by the prior formation of the hydride complex followed by the addition to it of the olefin, which is then reduced, rather than vice versa. This postulate is consistent with the observation that ethylene is reduced by the hydride complex, 10.2, while the ethylene complex, 10.3, is not readily reduced by hydrogen. The lower stability of the propene relative to the ethylene complex explains the ability of the system to act as a hydrogenation catalyst in the former but not in the latter case. When solutions of the rhodium complex are treated with a mixture of hydrogen and propene, the dihydride will be formed preferentially, whereas ethylene would form a more stable complex and prevent more than slow formation of the dihydride.

Proton magnetic resonance studies have shown the configuration of the dihydride complex to be either 10.4 or 10.5. The mechanism of the addition of the olefin to the hydride and its subsequent reduction is not yet clear. However, if a deuterium/hydrogen mixture is used during the reduction of 1-pentene there is essentially quantitative formation of the dideuteroparaffin and its non-deuterated analogue. It follows that the H_2 or D_2 molecule has been added to only one olefin and that there are no exchange reactions. Further, it has been shown that the addition takes place by a *cis* mechanism. It was therefore postulated that both the hydrogens add essentially simultaneously to the olefin; recent work, however, suggests that an alkyl intermediate may be involved followed by rapid transfer of hydrogen from the metal to yield alkane.

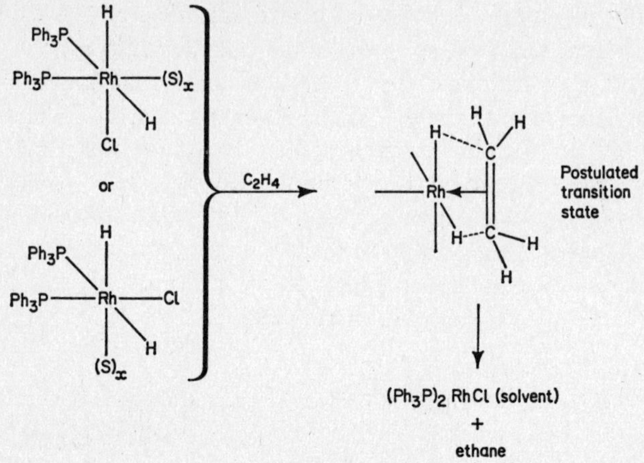

The oxidation of ethylene to acetaldehyde

Treatment of aqueous solutions of palladium(II) chloride with ethylene gives an ethylene-palladium complex which is readily hydrolyzed to form acetaldehyde and palladium metal. The palladium metal may be reoxidized by copper(II) chloride, in a continuous process:

$$C_2H_4 + PdCl_4{}^{2-}\,aq \longrightarrow [C_2H_4PdCl_3]^- \underset{}{\overset{H_2O}{\rightleftharpoons}}$$

$$[C_2H_4PdCl_2(H_2O)] \underset{H_3O^+}{\overset{H_2O}{\rightleftharpoons}} [C_2H_4PdCl_2(OH)]^- \quad \text{(complex decomposition inhibited by acid)}$$

$$[C_2H_4PdCl_2(OH)]^- \underset{slow}{\overset{H_2O}{\longrightarrow}} \left[\begin{array}{c} H\ H \\ | \ \ | \\ HO-C-C-PdCl_2(OH_2) \\ | \ \ | \\ H\ H \end{array} \right]$$

$$\Big\downarrow \text{fast}$$

$$CH_3CHO \longleftarrow HO-CH=CH_2 + Pd + 2Cl^- + H_3O^+$$

$$Pd + 2Cu^{2+} + 6Cl^- \longrightarrow [PdCl_4]^{2-} + 2\,CuCl$$

$$2\,CuCl + 2H^+ + \tfrac{1}{2}O_2 \longrightarrow 2Cu^{2+} + 2Cl^- + H_2O.$$

As shown above, the oxidation of the ethylene is thought to proceed via nucleophilic attack of a hydroxide ion on the ethylene. Palladium(II) chloride also catalyses the oxidation of propylene to acetone. Kinetic studies suggest that the rate-determining step is the addition of hydroxide to the co-ordinated olefin, viz:

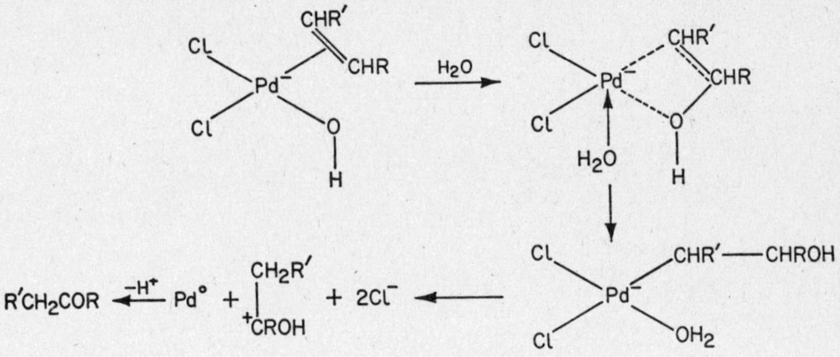

The oxidation of ethylene by palladium catalysts in acetic acid solution containing sodium acetate gives vinyl acetate. This reaction is the basis of the industrial preparation of vinyl acetate. The reaction may proceed as follows,

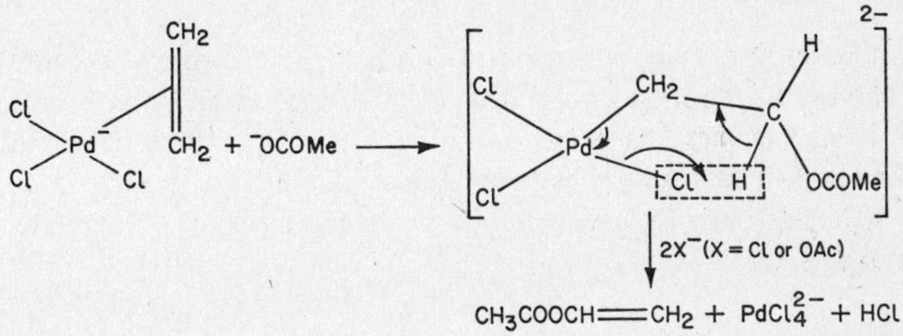

$$CH_3COOCH{=\!=}CH_2 + PdCl_4^{2-} + HCl$$

The trimerization of butadiene and related reactions catalyzed by some π-allyl complexes

Treatment of nickel acetylacetonate with aluminium alkyls in the presence of cyclododecatrienes or butadiene, or of bis-π-allyl nickel with butadiene, gives cyclododecatriene nickel, a volatile blood-red crystalline compound 10.6. This 16-electron nickel complex reacts catalytically and very rapidly with butadiene at 20° liberating isomers of cyclododecatriene. The major product is the *trans-trans-cis* isomer; small amounts of the *trans-cis-cis* isomer are also formed. If the reaction with butadiene is carried out at −40° the bis-π-enyl complex, 10.7, is isolated in which the ligand is the same as that found in the complex formed from butadiene and ruthenium salts (p 180). The carbonylation (Figure 53) provides evidence for the nature of complex, 10.7. The cyclization of the butadiene is believed to occur by internal electron shift processes, after the butadiene molecules have become attached to the metal.

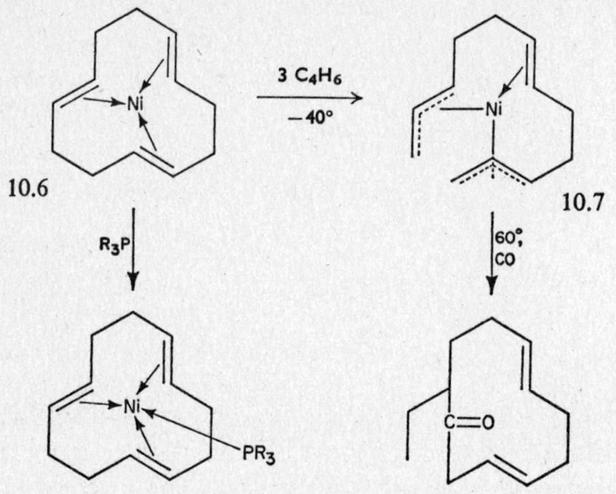

Figure 53. Some reactions of cyclododecatriene nickel.

If the complex, 10.6, is treated with a compound which acts as a non-labile ligand to the metal, then instead of trimerization only dimerization of butadiene occurs, since the attachment of a third butadiene molecule is prevented. The reaction is represented below. The intermediate complex, 10.8, has been isolated when Do = tris(2-biphenyl) phosphite.

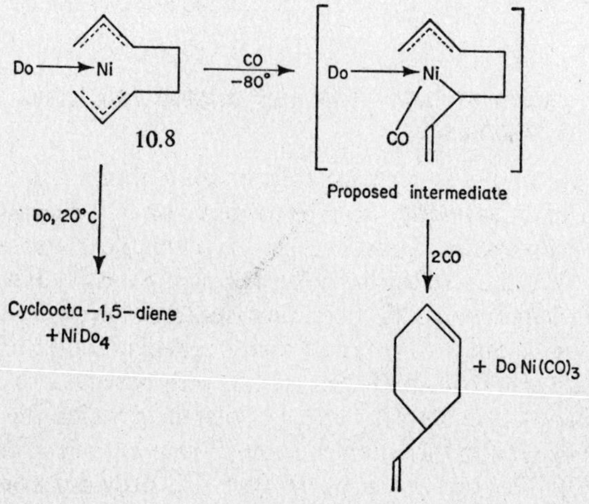

The reaction of the complex, 10.8, with carbon monoxide again illustrates how the course of the reaction may be changed when the nickel is complexed by the non-labile ligand. It can be seen that the choice of suitable ligands can in principle enable elegant control of catalytic syntheses. Such developments indicate that homogeneous catalysis has a very promising, even exciting, future.

Hydroformylation and related carbonylation reactions

As was mentioned earlier (p 189), an important property of transition metal σ-alkyls is their ability to undergo insertion reactions, especially with carbon monoxide, when acyl complexes are formed. Frequently such carbonylation reactions are reversible. Studies on the decarbonylation of acetylmanganese pentacarbonyl labelled with ^{14}C in the acetyl-CO group indicate that this carbonyl group is retained in the molecule as a carbon monoxide ligand:

$$Me^*COMn(CO)_5 \xrightarrow{\text{heat}} MeMn(CO)_4{}^*CO + CO$$

as no appreciable radioactivity was found in the gas phase. Similarly acetyl manganese pentacarbonyl formed by the carbonylation of methyl manganese pentacarbonyl with ^{14}CO possessed no activity in the acetyl group:

$$MeMn(CO)_5 + {}^*CO \xrightarrow{\text{pressure}} MeCOMn(CO)_4{}^*CO$$

Similar carbonylations can be effected by treatment of transition metal σ-alkyls with ligands other than carbon monoxide, e.g. triphenyl phosphine, phosphites, primary amines or iodide ion:

$$MeMn(CO)_5 + Ph_3P \longrightarrow MeCOMn(CO)_4PPh_3$$

$$MeMn(CO)_5 + LiI \longrightarrow Li^+[MeCOMn(CO)_4I]^-$$

The above experiments, together with kinetic and infrared studies indicate that carbonylation and decarbonylation are intramolecular processes. Two main mechanisms for the carbonylation can be envisaged— carbonyl insertion (a) or methyl migration (b):

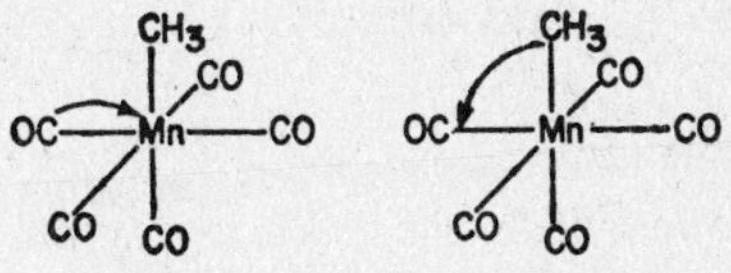

(a)   (b)

Infrared studies using ^{13}CO tracers are in agreement with the methyl migration mechanism (b) for both the carbonylation and decarbonylation processes. A similar mechanism has been proposed for the conversion of *trans*-$MeCOMn(CO)_4PPh_3$ into *cis*-$MeMn(CO)_4PPh_3$:

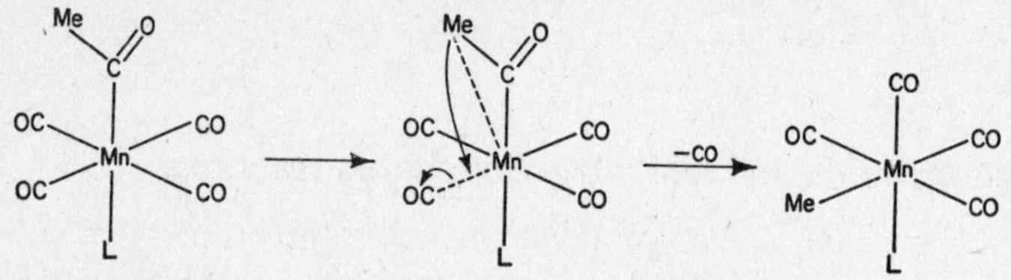

Many other σ-alkyl complexes behave similarly, although the ease of carbonylation (or decarbonylation of the acyl derivatives) varies quite widely with the nature of the complex. Thus $F_3CCOCo(CO)_4$ decarbonylates at 30°C whereas $F_3CCOCo(CO)_3PPh_3$ does not begin to lose carbon monoxide below about 130°C.

The conversion of olefins into aldehydes and ketones using carbon monoxide and hydrogen under pressure is of considerable industrial importance. The reactions involve corbonylation and decarbonylation processes. It has been shown that $HCo(CO)_4$ is the active catalyst in these systems. This hydride is formed under the high pressure (100 atm; 1:1 H_2:CO and temperature (100–300°) conditions used in hydroformylation reactions.

Ethyl cobalt tetracarbonyl is formed smoothly and reversibly under mild conditions from $HCo(CO)_4$ and ethylene. Infrared studies show that alkylcobalt tetracarbonyls are in equilibrium with the acyl tricarbonyl derivatives in solution at room temperature. Further evidence for these 16-electron intermediates comes from kinetic studies on the reduction of acetylcobalt tetracarbonyls by hydrogen or by $HCo(CO)_4$. In both cases the reduction is strongly inhibited by CO, suggesting that it is not acetylcobalt tetracarbonyl itself but the dissociated complex $RCOCo(CO)_3$ which is reacting:

(*a*) $RCOCo(CO)_4 \rightleftharpoons RCOCo(CO)_3 + CO$ (forward reaction repressed by carbon monoxide)

$RCOCo(CO)_3 \xrightarrow{HCo(CO)_4} RCHO + Co_2(CO)_7$; $\quad Co_2(CO)_7 \underset{-CO}{\overset{+CO}{\rightleftharpoons}} Co_2(CO)_8$

$RCOCo(CO)_3 \xrightarrow{H_2} RCHO + HCo(CO)_3$; $\quad HCo(CO)_3 \underset{-CO}{\overset{+CO}{\rightleftharpoons}} HCo(CO)_4$

On the basis of these equilibria the following formal mechanism for the hydroformylation of ethylene has been proposed:

$$HCo(CO)_4 + H_2C{=}CH_2 \rightleftharpoons C_2H_5Co(CO)_4 \rightleftharpoons$$

$$C_2H_5COCo(CO)_3 \xrightarrow[\text{hydrogenolysis}]{H_2 \text{ or } HCo(CO)_4} C_2H_5CHO + Co_2(CO)_8$$

The addition of $HCo(CO)_4$ [or $HCo(CO)_3$] to unsymmetrical olefins can occur in two directions, and both possible isomeric alkylcobalt tetra-carbonyls are often obtained. Moreover $HCo(CO)_4$ can isomerise olefins, and its subsequent addition to these isomerized olefins would give further isomeric alkyl cobalt tetracarbonyls. Therefore mixtures of aldehydes and ketones are often obtained under the conditions of the OXO reaction. The products are subsequently reduced to alcohols, the present production capacity for these being about 200 000 tons per year in Britain.

Catalytic conversion of acetylenes to αβ-unsaturated acids in the presence of nickel carbonyl: (Reppe process)

In inert organic solvents nickel carbonyl reacts with diphenylacetylene to form tetraphenylcyclopentadienone and bis-tetraphenylcyclopentadienone nickel (cf $Fe(CO)_5$ p 229). In the presence of aqueous acids (e.g. acetic, hydrochloric) acetylenes are converted into αβ-unsaturated acids, e.g. acetylene itself yields acrylic acid. With carbon monoxide under pressure, nickel carbonyl is continuously regenerated, so that the reaction becomes catalytic. The mechanism of this process is not yet understood, but it is found that water is essential. It is possible that formation of an inter-mediate σ-alkenyl nickel complex is involved, which affords the un-saturated acid after carbonylation and hydrolysis of the acyl derivative:

$$R{\cdot}C{\equiv}C{\cdot}R \xrightarrow{Ni(CO)_4} \begin{array}{c} R{\cdot}C{\equiv}C{\cdot}R \\ \downarrow \\ Ni \\ {/} | {\backslash} \\ C\ \ C\ \ C \\ O\ \ O\ \ O \end{array} \xrightarrow{HX} \begin{array}{l} RCH{=}CR'{-}Ni(CO)_2X \\ CO \downarrow \text{ (See p 241)} \\ RCH{=}CR'{-}CO{-}Ni(CO)_2X \\ H_2O \downarrow CO \\ RCH{=}CR'CO_2H + Ni(CO)_4 + HX \end{array}$$

Vitamin B₁₂ chemistry and related topics

Octahedral complexes derived from transition metal ions which have d^3 or low-spin d^6 electron configurations have considerably greater kinetic stability to ligand substitution or exchange than those formed by ions in other d-electron configurations. This stability can be associated in part with the spherical symmetry of the electron clouds in such configurations. As kinetic stability is essential for the isolation of σ-organocomplexes of the transition elements, it is not surprising that many stable complexes of the general formula ML_5R are known (where M is, for example, Cr^{III}, Co^{III}).

R

Treatment of chromium(II) salts in aqueous solution with 2-bromo-methylpyridine leads to air-stable solutions of the cation,10.9, from which crystalline salts may be isolated.

$$Cr^{2+}aq. \; + \qquad\qquad\qquad \longrightarrow \qquad\qquad [\qquad CH_2—Cr(H_2O)_5]^{2+}$$

10.9

Similarly complexes $[(CN)_5CoR]^{3-}$ may be prepared from the anion $[(CN)_5Co]^{3-}$ and organic halides, e.g.

$$[(CN)_5Co]^{3-} \begin{array}{c} \xrightarrow{\text{PhCH}_2\text{Br}} [(CN)_5Co—CH_2Ph]^{3-} \\ \xrightarrow{\text{H}_2\text{C}=\text{CH.CH}_2\text{Cl}} [(CN)_5Co—CH_2—CH=CH_2]^{3-} \end{array}$$

Aqueous solutions of the complexes $[(CN)_5CoR]^{3-}$ are stable in the absence of oxygen for weeks, but they are decomposed by acids. Some of them form nitriles on treatment with acid followed by base. It is suggested that the reactions proceed via intramolecular migration of the group R:

$$\underset{Co}{\overset{R}{|}}—C\equiv N \xrightarrow{H^+} \underset{Co}{\overset{R}{|}}—C\equiv \overset{+}{N}H \longrightarrow \overset{+}{Co}—\underset{}{\overset{R}{|}}C=NH \xrightarrow{OH^-} RCN + H_2O$$

This mechanism may be compared with that proposed for the acylation of alkyl-metal complexes (p 241).

A group of coenzymes (Vitamin B_{12} group) has been isolated from rat liver extract. These substances, which contain an octahedral Co(III) atom in the environment shown in Fig. 54(a), are cofactors in the promotion of growth in animals on diets containing vegetable proteins. They also act as anti-pernicious anaemia agents in humans. The crystal structure of Vitamin B_{12} coenzyme, where L = 5,6-dimethylbenzimidazole, determined by Hodgkin and her co-workers, shows that R is an adenine nucleoside group which is attached directly to cobalt by the 5′ carbon atom of a deoxyribose residue. Another coenzyme of this group—methyl-cobalamin, contains a cobalt-methyl bond. This substance is thought to act as a coenzyme in the conversion of homocysteine to methionine in biological systems.

Reduction of cyanocobalamin (Figure 54(a) R = CN) with sodium borohydride affords a diamagnetic grey-green product (Vitamin B_{12s}) which reacts with acyl or alkyl halides and with olefins to give cobalamin σ-organo derivatives. There has been some discussion whether Vitamin B_{12s} is a cobalt(I) anion or whether it is cobalamin hydride. Some of the

Homocysteine

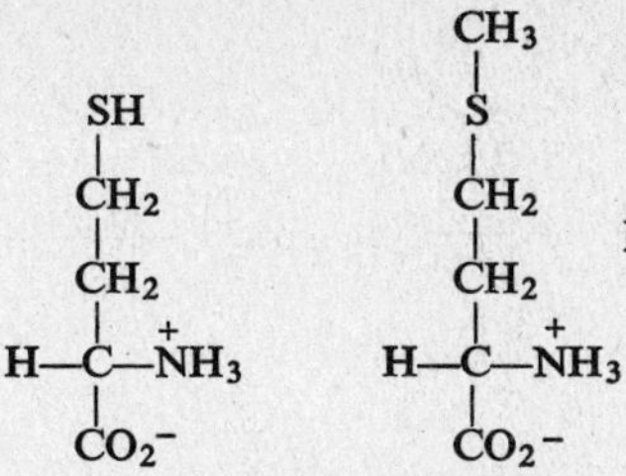

$$
\begin{array}{c}
\text{SH} \\
| \\
\text{CH}_2 \\
| \\
\text{CH}_2 \\
| \\
\text{H}-\text{C}-\overset{+}{\text{N}}\text{H}_3 \\
| \\
\text{CO}_2{}^-
\end{array}
$$

$$
\begin{array}{c}
\text{CH}_3 \\
| \\
\text{S} \\
| \\
\text{CH}_2 \\
| \\
\text{CH}_2 \\
| \\
\text{H}-\text{C}-\overset{+}{\text{N}}\text{H}_3 \\
| \\
\text{CO}_2{}^-
\end{array}
$$

Methionine

reactions shown in Figure 55 suggest at first sight that a Co—H bond is present whilst others are characteristic of anionic cobalt complexes. However, the reactions can be best understood in terms of an anionic complex.

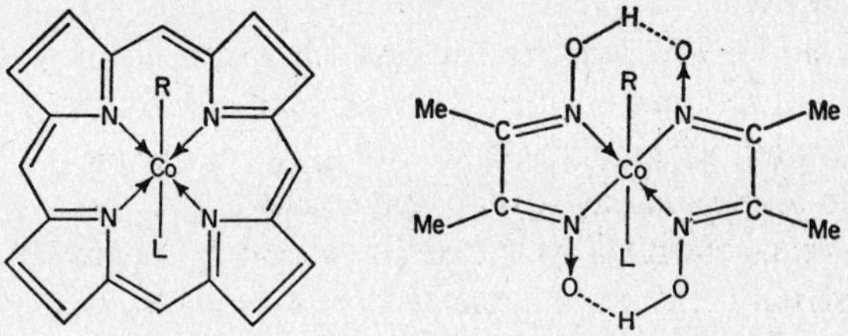

(a)   (b)

Figure 54. (a) A simplified representation of the structure of Vitamin B_{12} derivatives. (b) Cobaloxim—a bisdimethylglyoxime complex of cobalt.

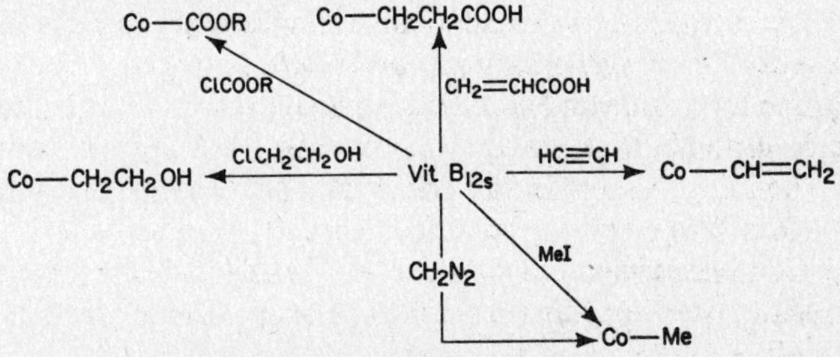

Figure 55. Some reactions of Vitamin B_{12s}.

A number of 'model' Vitamin B_{12} systems have been prepared and studied. The dimethylglyoxime complex, Figure 54(b), shows marked analogy in its chemical reactions to cyanocobalamin; for this reason it is called cyanocobaloxim (the complex is represented $Co(D_2H_2)CN.L$). Cyanocobaloxim is readily reduced and the product may be considered analogous to Vitamin B_{12s}. The preparation of cobaloxim-σ-organo derivatives from the reduced complex is shown below and may be compared with the reactions of Vitamin B_{12s}.

The nature of the reduced cobaloxim species in aqueous basic solution is

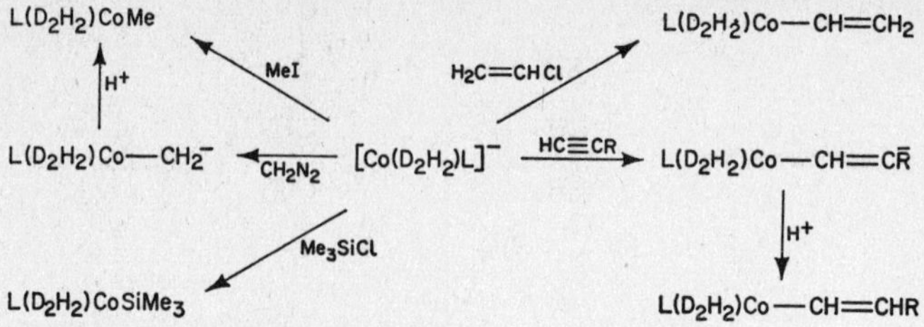
L(D2H2)CoMe
H+
MeI
H2C=CHCl
L(D2H2)Co—CH=CH2
L(D2H2)Co—CH2−
CH2N2
[Co(D2H2)L]−
HC≡CR
L(D2H2)Co—CH=CR−
Me3SiCl
H+
L(D2H2)CoSiMe3
L(D2H2)Co—CH=CHR

unknown but it is suggested that there may be both a hydride and an anionic complex in equilibrium, e.g.

$$H^+ + [Co(D_2H_2)L]^- \rightleftharpoons Co(D_2H_2)LH$$

Nitrogen fixation

It has been long known that certain bacteria which occur in the nodules of leguminous plants react with molecular nitrogen forming nitrogenous products, including ammonia. Other biological agents which fix nitrogen are blue-green algae such as *Nostoc* and yeasts.

For many years chemists have been intrigued by the contrast between these ambient temperature enzyme reactions and the great difficulty in getting molecular nitrogen to undergo chemical reactions under mild conditions in the laboratory. Only lithium and, to a lesser extent, a few other metals react slowly at room temperature.

Within the last few years, however, many research groups have tackled the problem of nitrogen fixation and some striking advances have been made.

(*a*) *Complexes of nitrogen with transition metals.* Nitrogen is isoelectronic with carbon monoxide and acetylene. It would therefore be expected on symmetry grounds that nitrogen could act as a ligand to transition metals, bonding either in a linear manner M—N≡N (compare M—C≡O, p 186) or by a π-bond (compare the acetylene-metal π-bond p 226). Complexes containing linear M—N≡N systems have indeed been prepared starting either from molecular nitrogen or from nitrogen compounds, e.g.

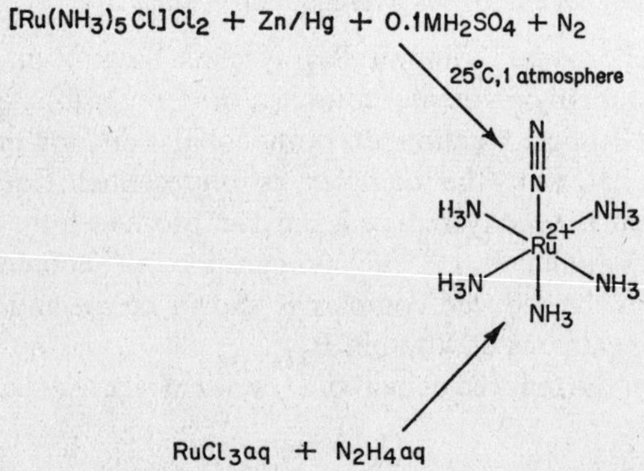
[Ru(NH3)5Cl]Cl2 + Zn/Hg + 0.1MH2SO4 + N2
25°C, 1 atmosphere
N
N
H3N
NH3
Ru 2+
H3N
NH3
NH3
RuCl3aq + N2H4aq

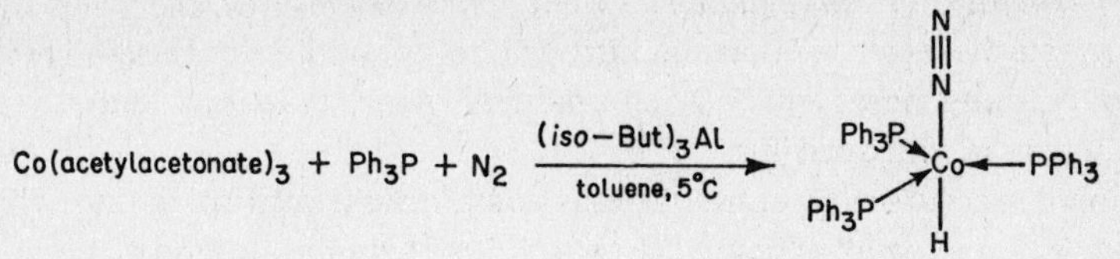

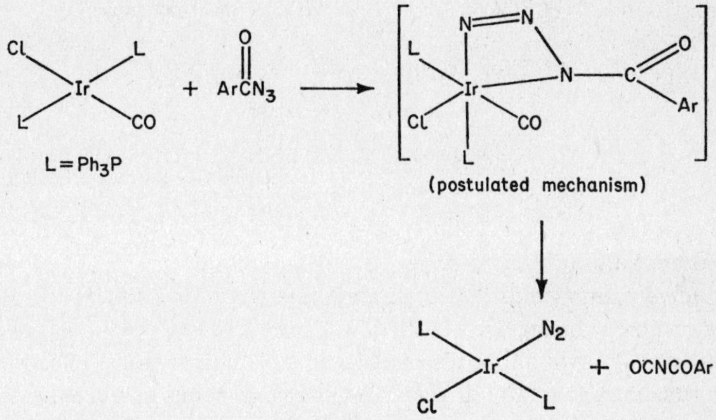

Co(acetylacetonate)$_3$ + Ph$_3$P + N$_2$ $\xrightarrow[\text{toluene, 5°C}]{(iso-But)_3 Al}$

$$\text{Ph}_3\text{P} \underset{\text{Ph}_3\text{P}}{\overset{\text{N} \equiv\equiv \text{N}}{\underset{|}{\overset{|}{\text{Co}}}}} \text{PPh}_3$$

L = Ph$_3$P

(postulated mechanism)

+ OCNCOAr

The crystal structures of the complexes $[(NH_3)_5 RuN_2]^{2+}$ and $(PPh_3)_3 CoN_2 H$ show that the M—N≡N systems, in these cases are essentially linear; N≡N is 1.12 and 1.16 Å respectively.

In these nitrogen complexes the N≡N stretching frequencies occur about 200 cm^{-1} lower than the N≡N stretch in nitrogen (2330 cm^{-1}). This may be attributed to the lowering of bond order in the N≡N bond by back-donations from the filled metal d-orbitals into the empty anti-bonding π^* orbitals of the nitrogen.

It is very interesting that the aquo cation $[(NH_3)_5 RuOH_2]^{2+}$ reacts with the ruthenium nitrogen compound forming a binuclear complex which has a linear, bridging-nitrogen, e.g.

$[(NH_3)_5RuN_2]^{2+}+[(NH_3)_5RuOH_2]^{2+} \rightleftharpoons [(NH_3)_5Ru—N\equiv N—Ru(NH_3)_5]^{4+}+H_2O$

It can be seen that, contrary to previous ideas, nitrogen can react very readily to give complexes. With suitable transition metal complexes nitrogen is *not* inert and behaves in a similar manner to carbon monoxide. For example, the reaction in toluene

$H_3Co(PPh_3)_3+N_2 \rightleftharpoons N_2HCo(PPh_3)_3+H_2$

proceeds at room temperature and atmospheric pressure and equilibrium is established within an half an hour.

(b) *Formation of ammonia under mild conditions.* So far, there has been no conclusive evidence that the nitrogen compound described above will react forming other nitrogen compounds. Most reactions lead to the evolution of nitrogen gas.

There are, however, some systems containing transition metals which react with nitrogen under mild conditions forming ammonia, e.g.

$$(\pi-C_5H_5)_2TiCl_2 + EtMgBr + N_2 \xrightarrow[25°C]{\text{Ether}} NH_3(0\cdot7 \text{ mole per mole Ti}) + \text{other products.}$$

$$TiCl_4 + K^+[t\text{-Butoxide}]^- + K_{metal} + N_2$$
$$\searrow \text{diglyme}$$

NH_3 (10–15% yield based on titanium) and other products.

$$FeCl_3 + EtMgBr + N_2 \longrightarrow NH_3 \text{ (trace quantities)}$$

Another titanium alkoxide system using sodium-naphthalenide as the reducing agent yields 340% of ammonia (based on titanium). Sodium is consumed during the reaction.

Little is known about the mechanisms of the reactions. In some cases intermediates containing metal-hydrogen systems may be important. It is clear that low oxidation states of the metals are required and it is highly probable that in this state the metals react with molecular nitrogen forming metal-nitrogen complexes as intermediates.

(c) *Reactions of some metal complexes which contain organo-nitrogen ligands.* There are known complexes which contain systems of the general type M—N—N. They readily gain or lose hydrogen or hydrogen ions. Whilst there is insufficient evidence to discern the mechanism of fixation of nitrogen to ammonia by biological or laboratory reactions, these systems provide *models* for possible mechanisms.

An example is:

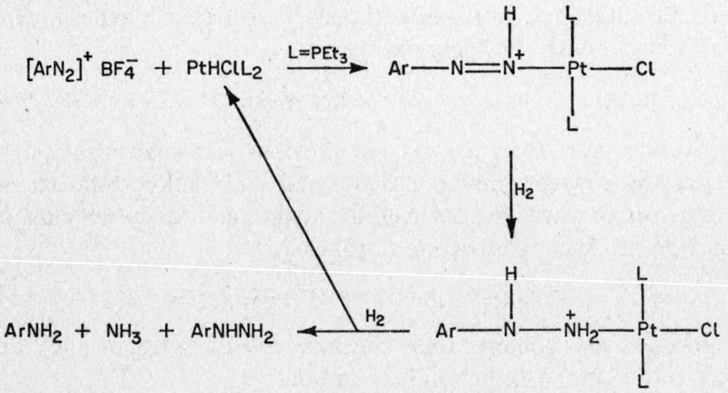

In the context of nitrogen fixation it is interesting to note that there are biological systems which fix methane. It seems probable that inorganic complexes which will readily react with methane under mild conditions will be found and these might well be of industrial importance.

BIBLIOGRAPHY

Transition metal hydrides
J. Chatt, *Proc. Chem. Soc.* 1962, 318; M. L. H. Green, *Endeavour* **26**, 1967, No. 99. Useful, brief reviews, suitable for undergraduates.
A. P. Ginsberg in *'Transition Metal Chemistry'*, Volume I, edited by R. L. Carlin (Edward Arnold, London, 1966); M. L. H. Green and D. J. Jones, *Advances in Inorganic Chemistry and Radiochemistry* **7**, 1965, 115. More detailed accounts.

Homogeneous catalysis
J. A. Osborn, *Endeavour* **26**, 1967, No. 99. A good review, suitable for undergraduates.
Advances in Chemistry Series, No. 70 (American Chemical Society), 1968; *Discussions of the Faraday Society*, No. 46, 1969. Accounts of recent research.

Palladium catalyzed reactions of olefins
A. Aguilo, *Advances in Organometallic Chemistry* **5**, 1967, 321. Oxidation of olefins.
J. Tsuji, *Accounts of Chemical Research* **1**, 1968, 144. Carbon-carbon bond formation via palladium complexes.

Carbonylation reactions and related topics
C. W. Bird, *Chem. Rev.* **62**, 1962, 283.
R. F. Heck, *Advances in Organometallic Chemistry* **4**, 1966, 243. Synthesis and reactions of alkyl- and acylcobalt tetracarbonyls.
A. J. Chalk and J. F. Harrod, *Advances in Organometallic Chemistry*, **6**, 1968, 119. Catalysis by cobalt carbonyls.

Oligomerisation of butadiene using transition metal complexes
G. Wilke, *Angew. Chem.* (international edition) **2**, 1963, 105; **5**, 1966, 151.

π-Allylnickel intermediates in organic synthesis.
P. Heimbach, P. W. Jolly and G. Wilke, *Advances in Organometallic Chemistry*, **8**, 1970, 29.

Nitrogen complexes
R. Murray and D. C. Smith, *Coordination Chemistry Reviews* **3**, 1968, 429. The activation of molecular nitrogen, including biological aspects.
A. D. Allen and F. Bottomley, *Accounts of Chemical Research* **1**, 1968, 360.
G. Henrici-Olivé and S. Olivé, *Angew. Chem.*, (international edition) **8**, 1969, 650. Useful, brief reviews.

Index

Organometallic compounds are indexed under the metal concerned. Where possible a major reference is indicated in **bold** type when several are given.